Electronic me
instrumentatio

Electronic measurement and instrumentation

K.B. KLAASSEN

IBM Almaden Research Center, San Jose

Translation from Dutch: S.M. Gee

CAMBRIDGE
UNIVERSITY PRESS

Published by the Press Syndicate of the University of Cambridge
The Pitt Building, Trumpington Street, Cambridge CB2 1RP
40 West 20th Street, New York, NY 10011-4211, USA
10 Stamford Road, Oakleigh, Melbourne 3166, Australia

Originally published in Dutch as Electrotechnisch Meten by VSSD and © VSSD 1983
© English translation Cambridge University Press 1996

First published in English by Cambridge University Press, 1996 as *Electronic measurement and instrumentation*

Printed in Great Britain at the University Press, Cambridge

VSSD

Contents

Preface

vii Preface

Our ability to make accurate measurements is one of the fundamental abilities enabling us to engage in science and engineering. Phrasing it differently: 'In physics only that exists which can be measured' (Max Planck). Even our daily life revolves around our ability to quantify delivered or consumed amounts (electrical energy), to verify desirable amounts (blood pressure) and safe amounts (alarm level) and to monitor and control amounts (traffic).

Measurement has a long history. Four thousand years before Christ, the Babylonians and the Egyptians were already conducting astronomical measurements. However trivial it may look today, no small achievement in measurement was the world-wide acceptance of one system for weights and measures (the SI-system).

Progress in electrical engineering, more particularly, the tremendous advance in electronics, has meant that virtually all measurements today are made in the 'electrical domain'. To measure a non-electrical quantity, a transducer is used to convert the measurement instrumentation from the respective non-electrical domain into an electrical signal.

The purpose of this book is to treat the fundamentals of measurement and instrumentation in such a way that, apart from providing the necessary elementary knowledge, insight into measurement is stimulated. The goal thereby is to enable the students or engineers using this textbook to independently solve their measurement problems independently. The level of presentation is chosen such that only knowledge of the most basic electrical engineering concepts and a familiarity with rather elementary mathematics are needed.

In this book the well-known approach of describing measurement instruments or devices was dropped in favour of a more 'system-orientated' approach. After the introduction in Chapter 1 of the questions of what constitutes a measurement and why we measure, the foundation for all measurements (measurement theory) is laid. Chapter 2 begins by showing that for the hierarchically highest-ranking measurements (cardinal measurements), a system of units is needed. It then shows that to conduct optimal measurements (i.e. measurements attaining the required accuracy with the least effort and cost, in the shortest time) it is essential that one is familiar with a number of alternative measurement methods. A measurement can only produce a finite certainty about the true value of the measurand; there are always measurement errors. An error analysis is therefore given in Chapter 2 that describes types of error, error propagation and causes of errors. The physical means for conducting a measurement, the measurement system, has a certain topology containing a number of different functions: transduction, signal conditioning and processing, display and registration of the measured quantity. These measurement functions are described in Chapter 3. In the next chapter several completely electronic measurement functions are discussed such as those for frequency, phase, voltage, etc. Much attention is given here to fully automated measurements using a computer: multiplexing, aliasing, sampling, analogue-to-digital conversion, data busses, etc. The appendix deals with the SI-system, conversion factors and some other topics useful to those studying measurement engineering.

This textbook found its origin in the lecture notes of the author when he was teaching an undergraduate course in measurement and instrumentation as a professor at the Department of Electrical Engineering of the Delft University of Technology in the Netherlands. Originally in Dutch, published by the Delft Student Press (VSSD), the manuscript was translated into English by S. M. Gee.

May, 1995,
San Jose, California, USA   K.B. Klaassen

1

Basic principles of measurement

The contents of this book on measurement and its applications can be divided into three sections: a general section (Chapter 1), a section in which we restrict ourselves to the measuring of physical quantities (Chapter 2) and a section in which we restrict ourselves even further to electrical and electronic measurements (Chapters 3 and 4). The latter restriction is not as drastic as it may seem; nowadays most measurements, including those concerning other than electrical quantities, are performed electronically. This necessitates conversion of the respective non-electrical quantity into a measurable electrical quantity. This conversion is accomplished by a so-called transducer (Chapter 3).

In this first chapter we will address the basics of measurement; we will define what we mean by measurement and reveal the purpose of measurement. We will further deal with the scientific underpinnings of this field (measurement theory) and, finally, we will study the question of why measurement of non-physical quantities is so difficult.

1.1 Definition of measurement

A possible operational description of the term measurement which agrees with our intuition is the following: 'measurement is the *acquisition of information*'. The aspect of gathering information is one of the most essential aspects of measurement; measurements are conducted to learn about the object of measurement; the *measurand*. This means that a measurement must be *descriptive* with regard to that state or that phenomenon in the world around us which we are measuring. There must be a relationship between this state or phenomenon and the measurement result. Although the aspect of acquiring information is elementary, it is merely a necessary and not a sufficient aspect of measurement: when one reads a textbook, one gathers information, but one does not perform a measurement.

A second aspect of measurement is that it must be *selective*. It may only provide information about what we wish to measure (the measurand) and not about any other of the many states or phenomena around us. This aspect too is a necessary but not sufficient aspect of measurement. Admiring a painting inside an otherwise empty room will provide information about only the painting, but does not constitute a measurement.

A third and necessary aspect of measurement is that it must be *objective*. The outcome of the measurement must be independent of an arbitrary observer. Each observer must extract the same information from the measurement and must come to the same conclusion. This, however, is almost impossible for an observer who uses only his/her senses. Observations made with our senses are highly subjective. Our sense of temperature, for example, depends strongly on any sensation of hot or cold preceding the measurement. This is demonstrated by trying to determine the temperature of a jug of water by hand. If the hand is first dipped in cold water, the water in the jug will feel relatively warm, whereas if the hand is first dipped in warm water, the water in the jug will feel relatively cold. Besides the subjectivity of our observations, we human observers are also handicapped by the fact that there are many states or phenomena in the real world around us which we cannot observe at all (e.g. magnetic fields), or only poorly (e.g. extremely low temperatures or high-speed movement). In order to guarantee the objectivity of a measurement we must therefore use artefacts (tools or instruments). The task of these instruments is to convert the state or phenomenon under observation into a different state or phenomenon that cannot be misinterpreted by an observer. In other words, the instrument converts the initial observation into a representation that all observers can observe and will agree on. For the measurement instrument's output, therefore, objectively observable output such as numbers on an alpha-numerical display should be used rather than subjective assessment of such things as colour, etc.. Designing such instruments, which are referred to as *measurement systems*, is the field of *(measurement) instrumentation*.

In the following, we will define measurement as the acquisition of *information* in the form of *measurement results*, concerning characteristics, states or phenomena (the measurand) of the world that surrounds us, observed with the aid of *measurement systems (instruments)*. The measurement system in this context must guarantee the required descriptiveness, the selectivity and the objectivity of the measurement. We can distinguish two types of information: information on the state, structure or nature of a certain characteristic, so-called *structural information*, and information on the magnitude, amplitude or intensity of a certain characteristic, so-called *metric information*. The acquisition of structural information is called a *qualitative measurement*, the acquisition of metric information is called a *quantitative measurement*. If the nature of the characteristic to be measured is not (yet) known, it must be determined first by means of a qualitative measurement. This can then be followed by a quantitative measurement of the magnitude of the respective characteristic.

If, for example, we wish to measure the sense of direction of a homing pigeon, we must first determine which physical quantity the pigeon uses to find its bearings: sunlight, the stars, the earth's magnetic field or maybe even the earth's gravitational field. Only then can we proceed to conduct a quantitative measurement of the sensitivity of this sense of direction.

Qualitative measurements determine the nature of that the quantity to be measured and provide us with the information we need to choose the devices for a quantitative measurement. In almost all cases, though, we already have this structural information and therefore need only to perform a quantitative measurement.

1.2 Why measuring?

Why is there so much measuring going on? Apparently to provide information about the world that surrounds us, the observers. One reason may therefore be that we wish to add to and to improve our perception of that world. Abstractly speaking, our aim is to increase our knowledge of the surrounding world and the relationships that exist between characteristics, states and phenomena in this world. This is the case even when we measure such day to day quantities as tire pressure, body temperature (fever), etc. The gathered information enables us to reduce seemingly complex characteristics, states, phenomena and relationships to simpler laws and relations. Thus, we can form a better, more coherent and objective picture of the world, based on the information measurement provides. In other words, the information allows us to create models of (parts of) the world and formulate laws and theorems. We must then determine (again by measuring) whether these models, hypotheses, theorems and laws are a valid representation of the world. This is done by performing tests (measurements) to compare the theory with reality. We have actually described the application of measurement in the 'pure' sciences. We assume that 'pure' science has the sole purpose of *describing* the world around us and is therefore responsible for our perception of the world. Fig. 1.1 illustrates this schematically.

In Fig. 1.1 the role of measurements in 'applied' sciences is also indicated. We consider 'applied' science as science intended to *change* the world. Thereto, it will use the models, laws and theorems of 'pure' science to modify the world around us. In this context, the purpose of measurement is to regulate, control or alter the surrounding world, directly or indirectly, based on the results of measurements and (existing) models, laws and theorems. The results of this regulating control can then be tested and compared to the desired results and any further corrections can be made.

Even a relatively simple measurement such as checking the tire pressure of a car can be described in the above terms: We have a hypothesis; we fear that the tire pressure is too low (or too high), otherwise we would not check the pressure. After measuring, we verify if the pressure, indeed, lies within the manufacturer's specified range. If not, we alter the pressure and measure it again, until it is correct.

We see from this that measurements form the essential link between the empirical world on the one hand and our abstraction thereof on the other. Our concept of the world around us is kept from becoming a dream concept

Figure 1.1. Measurement as the link between the real world on the one hand and its concept in the 'pure' sciences (p.s.) and 'applied' sciences (a.s.) on the other.

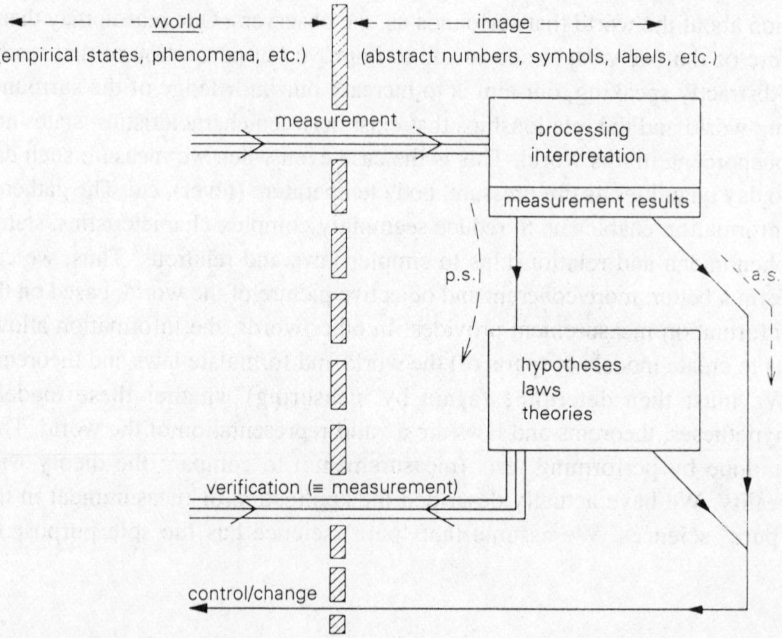

by measurements, forming the link between reality and perception. Without measurement we would only have a philosophical concept (as the ancient Greeks did). Also, faced with a number of different concepts, we would not be able to test which one was valid. Each individual could maintain a personal perception, without the possibility of ever achieving generally accepted concepts, as is the case now. Therefore, without measurement, our society could have got stuck in the time of alchemists, astrologers and sorcerers!

1.3 Measurement theory

In the previous section we have seen that measurements form the essential link between the empirical world and our theoretical, abstract image of the world. This concept forms the basis of a theory of measurement. In this theory a measurement result is considered to be a *representation* of the actual empirical quantity. Measurement theory treats measurements as a *mapping* of elements of a *source set* belonging to the empirical domain space (see Fig. 1.2) onto the elements of an *image* (or outcome) *set* which is part of the abstract range (or image) space. The quantity to be measured (the *measurand*) is an element of the source set. For instance, in the electrical domain we measure electrical current (source set) but only within a certain range of magnitude (elements). The result of the measurement process is abstract; it forms an element of the image set in the abstract range space. For example, the magnitude of the electrical current to be measured in the above example is (by measurement) assigned a certain number (element) out of the set of real numbers (image set). In other words, the elements of the source set are empirical characteristics of states and phenomena of the world around us; the elements of the image set are symbols of the abstract image set of symbols. The symbols can be numbers (quantitative measurements) but can also be, for example, names (quantitative measurements).

Restricting the definition of measurement further, measurement theory states that measurement is the mapping of elements from an empirical source set onto elements of an abstract image set *according to a particular transformation function*. The transformation function consists of the assignment algorithms, rules or procedures that define the representation of empirical quantities by abstract symbols. In practice the assignment algorithm, rule or procedure is implemented by the employed measurement system. The measurement system therefore determines the representation. As stated earlier, this representation must be done in a descriptive, objective and selective way. Thus, the image set must consist of elements (measurement outcomes) which are abstract symbols with a unique meaning about which, by definition, all observers agree.

A measurement must be descriptive. In measurement theory this is described in terms of set theory: the relations that exist between the elements of the source set must be maintained under the transformation in the image set, for example, 'larger than', 'equal to' and 'smaller than'. The set of relations between the elements of the source set is referred to as the *relational system* (of the source set).

This empirical relational system determines the *structure* of the source set. Likewise, an abstract relational system determines the structure of the image set (for instance, the set of relations that apply to the set of integer numbers). A measurement (representation) is now called *descriptive* if the relational system or structure of the empirical source set is invariant under the transformation (measurement). The measurement only represents that which is measured if the two relational systems are identical; otherwise information is lost in the mapping. An example is measuring with a very low resolution; two different current magnitudes are mapped onto the same outcome, and are indistinguishable from each other.

Let us now try to put this more formally: Assume that an empirical source set S consists of n elements s_i, so that $S = \{s_1, s_2, ..., s_n\}$. Let there exist k empirical relations R_j between the elements $s_i \in S$, so $R_j \subset E^{nj}$. Further, let the abstract image set I consist of m elements i_i, $I = \{i_1, i_2, ..., i_m\}$. Between these elements there are l relations N_j, so that $N_j \subset I^{mj}$. Clearly, if $k \neq l$, for instance, if l is larger than k, the measurement outcome will suggest more information than is actually present in the measurand. Similarly, if the number of elements of the two sets is not equal ($m \neq n$), for instance, $n > m$, the resolution of the mapping process may be inadequate. Therefore, for simplicity, let us assume $k = l$ and $m = n$. Now, let there be a function f that maps the elements of S onto I. We have to assume that this function is a single-valued, monotonic function in s_i. This will ensure a unique mapping

Figure 1.2. Measurement constitutes according to measurement theory the mapping between an empirical domain and a range or image space.

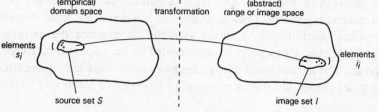

(empirical) domain space

transformation

(abstract) range or image space

elements s_i

elements i_i

source set S

image set I

onto I. The two relational systems $\langle S, R_1, R_2,...,R_k \rangle$ and $\langle I, N_1, N_2,...,N_l \rangle$ are *isomorphic* if:

$$\langle s_1, s_2,...,s_{nj} \rangle \in R_j \Leftrightarrow \langle f(s_1), f(s_2),...f(s_{nj}) \rangle \in N_j$$

Again, the function f is implemented by the employed measurement system in the form of an algorithm, rule or procedure.

If we require isomorphism, we ensure that the relations between the elements of S are preserved; the structure remains the same. In other words, the information contained in the relations is not lost during the measurement. Although the requirement for isomorphism may preserve the *structure* of S, it does not define the representation entirely. It allows a certain amount of freedom in choosing the measurement procedure. This freedom can be demonstrated by transforming the measurement results using 'legal' transformations into new results which contain the same information. The requirement for isomorphism does not yield a single, unique representation, but rather a group of congruent representations. The results obtained with each of these representations can be transformed into one another, without loss of information. Such *allowed transformations* do not affect the structure of the empirical domain S. The information contained in the measurement results is invariant with respect to the allowed transformations. The allowed transformations therefore show exactly how unique the assignment of measurement values is. If we were to contemplate a more detailed assignment of values (fewer allowable transformations, more unique measurement results), we would come across two limitations, a theoretical one and a practical one. The *theoretical limit* is the relational system that actually exists in the empirical domain S. We cannot map S onto an abstract image I meaningfully, if there are more relations in A than there are in S. This would add non-existing information to the measurement. The *practical limit* is the mapping which the available physical measurement system can accomplish. Some instruments have, for instance, a low resolution, others a very limited bandwidth. The theoretical limit is obviously the upper limit for assigning measurement outcomes: no matter how sophisticated the instrument used may be, it should never assign an outcome with more structure than is actually present in S.

We can therefore categorise measurements as indicated in Table 1.1 (third column), according to the allowable transformations $g(a)$ of the measurement results a. Hereby, we are taking into account both the practical and the theoretical limitations discussed above.

We can also categorise measurements according to the relations that actually exist in the empirical domain S. Here we, of course, take only the theoretical limitations into account (see Table 1.1, second column). If we were to deal with ideal instruments, the entries in columns two and three in Table 1.1 would be the same. In this case, the second column describes the actual relations between the elements s_i of S. The third column describes the allowable transformations on the elements of $s_i \in S$ and on the elements $i_i \in I$ of the image space. An ideal image results when a measurement is performed with instruments which at least map the relations of the second column onto I (requirement for isomorphism).

In this way we can distinguish five types of measurement: nominal, ordinal, interval, ratio and cardinal. Each successive measurement in this list is of a higher type and has all the characteristics of the preceding, lower measurements. We have already seen that the measurement system can, at most, establish the empirical relations which actually exist between the elements of the empirical domain S. In that case the measurement is ideal. In practice, the purpose of the measurement determines whether or not a lower measurement will suffice. It should be noted that in a measurement system consisting of several, serially circuited subsystems, none of the subsystems may destroy the desired information in the final result. Therefore, the type of measurement required for the total measurement system is also required for each subsystem.

The above stated requirement of isomorphism solves the so-called *representation problem* of the measurement theory. If our instrumentation

Table 1.1. *Information contained in the results of various types of measurements.*

Measurement	If s is to be measured, the measurement system must be able to determine:	Allowable transformations $g(i)$ of results i must satisfy:
nominal	$s_1 \not\equiv s_2$	$g(i_1) \not\equiv g(i_2)$ if $i_1 \not\equiv i_2$ one-to-one functions $g(i)$
ordinal	$s_1 \gtrless s_2$	$g(i_1) \gtrless g(i_2)$ if $i_1 \gtrless i_2$, monotonically increasing functions $g(i)$
interval	besides $s_1 \gtrless s_2$ also $|s_1 - s_2| \gtrless |s_3 - s_4|$	$g(i_1) \gtrless g(i_2)$ if $i_1 \gtrless i_2$ and $|g(i_1) - g(i_2)| \gtrless |g(i_3) - g(i_4)|$ if $|i_1 - i_2| \gtrless |i_3 - i_4|$ linearly, increasing function: $g(i) = mi + n, \; m > 0$
ratio	$s_1 \gtrless ms_2$, in which m is a rational number	$g(i_1) \gtrless mg(i_2)$ if $i_1 \gtrless mi_2$, thus all functions: $g(i) = mi, \; m > 0$
cardinal	$s = mU$, in which U is the unit	$g(i) \gtrless mU$ if $i \gtrless mU$, therefore identity function: $g(i) = i$

satisfies this requirement, the measurement result is representative of that which is to be measured; the result represents the measurable quantity. From Table 1.1 one can see, however, that the so-called *problem of uniqueness* has not yet been solved. For this we must also require that the domain E in question warrants a *cardinal measurement* and that the instruments are capable of such a measurement. A measurement is unique if and only if this is the case. The result of this kind of cardinal measurement can only be transformed in one way: by the identity transformation $g(i) = i$.

As we have seen above, the most elementary form of measurement is a *nominal measurement* (from Latin, *nomen* = name). In this measurement, the quantities which are to be measured are divided into a number of classes or groups, in such a way that the measurand fits into one and only one class or group. This procedure yields a classification measurement, for example, classifying an illness by diagnostic means, classifying the flora, fauna and so on. The result of a nominal measurement can be a value, but can also be a name or a symbol. Here a numerical value does not have the same weight as we are used to with higher measurements. It is merely a label. None of the arithmetic relations between numbers hold here. The only thing significant in a nominal measurement is that equal characteristics, states or phenomena are assigned the same labels, and that different characteristics are assigned different labels. The essence of this measurement is the absolute significance of equality and inequality. This leads to the conclusion that any one-to-one function with a one-to-one inverse function constitutes an allowable transformation for the results of a nominal measurement. A nominal measurement is the most primitive type of measurement. Every *qualitative measurement* (see Section 1.1) is a nominal measurement, as all we determine is whether the nature of the measured quality is equal or not equal to a known characteristic. An example of nominal measurement in the technical sciences is the class of measurements performed by *detection systems*. These systems have been deliberately designed with a simple binary output. A fire detection system signals 'no fire' below a certain temperature, and 'fire' above that temperature. Temperature is, of course, also a quantity that can be measured cardinally. The empirical relations in the thermal domain are far more extensive than a nominal measurement can represent. A nominal measurement is chosen for economic reasons; it is simple and cheap. Burglar alarms and smoke detectors are detection systems which make nominal measurements: the measurement results are images of events which are mutually exclusive. A nominal measurement cannot indicate which of the events or phenomena is larger or smaller. All that can be determined is 'does occur' or 'does not occur'. If there are more than two possible outcomes, a nominal measurement can indicate which one has occurred. For example, one can label the colour of an object with the name of the colour listed on the back of a colour chart or colour palette, which most resembles the object colour.

Other, higher types of measurement permit a more subtle assignment of values than just a classification. *Quantitative* measurement starts with an *ordinal measurement* (L. *ordinalis* = order). In such a measurement the relations 'larger than', 'equal' and 'smaller than' remain intact. The ordinal measurement is based on the principle of the absolute significance of the comparison of two quantities. We can now speak in terms of 'equal to', 'more than' or 'less than', based on such a measurement. With an ordinal measurement we can determine the *relative magnitude* of two characteristics; we can arrange characteristics according to size, magnitude or intensity. There are numerous examples of ordinal measurements: school exams, intelligence tests, arranging people according to strength by holding tug-of-war contests, and so on. Even Crozier's classification of women according to the care they take of their appearance (negligées — peu élégantes — standard moyen parisien — élégantes) is a type of ordinal scale! The values assigned by an ordinal measurement represent the relative order of magnitude, but do not, however, provide any further relations. For example, a person who has an IQ of 140 is not necessarily twice as intelligent as someone with an IQ of 70, or, the difference in acquired knowledge between a student with a grade A and a student with a grade B for a certain exam is not necessarily the same as that between someone with a grade E and someone with a grade F. It can easily be seen that we leave the relations 'smaller than', 'equal to' and 'larger than' intact when we apply a monotonically increasing function to the outcome of an ordinal measurement. This monotonic transformation is obviously an allowable transformation. We can therefore also conclude that calculating the average of several exam results is pointless, as we are allowed to transform the results of any particular exam with any monotonically increasing function. If we calculate the average again after such a transformation, we will find totally different results. Therefore, the average of an ordinal measurement does not contain any significant information.

The higher type of measurement is the so-called *interval measurement*. With an interval measurement we can determine not only whether one quantity is larger than, equal to or smaller than another, but also whether this is true for the interval (or difference) (see Table 1.1). The origin or zero of an interval measurement is not fixed, however (floating origin). Examples of interval measurements are time measurements and temperature measurements in degrees Celsius (using a mercury thermometer). The information in an interval measurement is not destroyed if we multiply the results with the same positive real number or if we add the same real number to all the results. The allowable transformations are all linear, increasing functions.

The next higher type of measurement has all the above mentioned characteristics plus a fixed origin. It is called *ratio measurement*. With this measurement the ratio of two quantities can be determined. The results of a ratio measurement may only be multiplied by a positive real number, as indicated in Table 1.1. For most physical quantities we can construct measurement instruments which can perform ratio measurements.

Finally, we come to the highest type of measurement, the *cardinal measurement*. Here, one relates the magnitude of a quantity to the magnitude of a predetermined (reference) quantity. The choice of the size, magnitude or intensity of the reference is arbitrary, provided this exact same reference is used in all cardinal measurements of the quantity. Fortunately, after much historical jumble, mankind has accepted a unit of reference for almost all physical quantities. For instance, a mass of 10 kg is 10 times the (internationally) agreed upon unit reference of 1 kg. It appears as if the result of a cardinal measurement (e.g. 10 kg) is no longer the ratio of the measured value and the reference value. Since there is no freedom left in a cardinal measurement, the results are fixed, see Table 1.1, and only the identity transformation is permitted. The results could therefore also be represented as just numbers, leaving out the reference symbol. In the example given above the mass would then be 10, instead of 10 kg. (Cardinal originates from the Latin *cardo* = that on which something hinges.)

Strictly speaking, when measuring, the only thing from which one may extract information is *the observation*. Unfortunately, the measured result or observation is not quite as perfect as we might wish, due to practical and also fundamental limitations. The image or outcome $i = f(s)$ provided by the instrumentation of the measurand s is a little vague. One cause of this vagueness is the thermal noise that is always present in every instrument. This means that we can actually observe very peculiar results, for example, in the case of a nominal measurement: $s_1 = s_2$ and, simultaneously, $s_1 \neq s_2$. This uncertainty obviously occurs when s_1 is almost equal to s_2 (although under a nominal measurement we cannot really speak of 'almost equal'). The resolution and repeatability of our measurement system are clearly not high enough. This is a fundamental problem; if we could make a system with a higher resolution the same problem will occur again, but now for two values closer together. Ergo, as we strive to bring about an abrupt cut in the otherwise continuous transition from 'not equal' to 'equal', and our only source of information is our observation, we must introduce an axiom:

− The cases $s_1 = s_2$ and $s_1 \neq s_2$ are mutually exclusive (correctness axiom).

If we are now confronted with the above mentioned uncertainty, we have to conclude that no statement can be made about the measurement result; it is inconclusive. Without the given axiom the conclusion would have been that s_1 and s_2 are both equal and not equal, as only the observed measurement result counts. In this manner measurement theory constructs a formally correct system, based on observations, with the aid of a number of axioms.

In the case of ordinal measurements there are several practical imperfections which would lead to incorrect statements based solely on the observation. This is prevented by postulating the following three axioms.

− If $s_1 < s_2$ then a finite real number n exists, so that $ns_1 \geq s_2$.

This means that s_1 cannot be infinitely small; due to the always present uncertainty in the measurement we simply cannot draw any conclusions on such a result.

− The cases $s_1 \geq s_2$ and $s_2 > s_1$ are mutually exclusive (correctness axiom).

− If $s_1 \geq s_2$ and $s_2 \geq s_3$ then $s_1 \geq s_3$. (transitiveness axiom).

We must recognise the above axioms, because when s_1 and s_2 are very close together, we can no longer distinguish between $s_1 > s_2$, $s_1 = s_2$ and $s_1 < s_2$. In order to arrive at a formal, logically correct system, in spite of the practical shortcomings of our physical measurement means, we have to complement our observations with the above axioms. In the case of an interval measurement we need two additional axioms, the correctness axiom and the transitiveness axiom applied to the intervals $|s_1 - s_2|$ and $|s_3 - s_4|$. The second axiom states that it is not possible to measure infinitely small intervals of quantities due to practical uncertainties. This way we distinguish between pure arithmetic, in which numbers and algebraic expressions are exact, and applied arithmetic, in which we work with approximated (measured) and therefore rational numbers. Measurement theory proves that even the uncertain measured numbers can be processed in a 'pure' manner, as long as the techniques discussed in the section on error propagation are used to determine the significance of a measurement result.

1.4 Measurement of non-physical quantities

Non-physical quantities found in non-technical fields are very difficult or even impossible to measure, mainly because they are part of very *complex systems* (people, organisations, societies, etc.).Cardinal measurements can

therefore very rarely be made when dealing with the non-physical characteristics, states or phenomena in such systems.

When measuring complex systems or objects, the measurand often also
depends on all kinds of other quantities. Usually, the nature and magnitude
of these dependencies are unknown. Furthermore, these parasitic quantities,
which influence the measurement result, are not constant during the measurement, making it impossible to correct any errors in the measurement.
This means that the measurement is no longer selective; other factors are
also being measured. Non-physical quantities are usually a part of a living
organism or an organisation of living beings. It is essential to the nature of
an organism or an organisation that they maintain (social, cultural, political)
interactions with their environment. For this reason, it is usually not possible
to perform an *isolated measurement* in the same way we can with inanimate
things (putting in a thermostat, hooking up to a fixed supply voltage, etc.). It
is, for example, not very practicable to isolate a section of the population
from the rest of the world, for the sake of an economic measurement
conducted to verify the relationship between the scarcity and the price of
consumer goods. The object of the measurement continues to interact with
its surroundings, to an extent and in a manner that remains unknown. These
interactions obscure and corrupt the measurement results to some unknown
degree, depending on the sensitivity to these interferences.

Also, the *repetition of an experiment* is often not possible when dealing
with such complex measurement objects. With physical measurements we
often repeat a measurement to see how reproducible it is and to determine
the magnitude of random errors. With non-physical measurements, however,
the object often behaves differently the second time, because it has either
learned from the first measurement, or it is no longer motivated to co-
operate, has become tired, etc. Think, for example, of taking the same
examination twice.

For ethical, political or financial reasons it is also usually not possible to
freely modify variables to determine how they influence the measurement
object, for example, increasing the scarcity of food in an economic
community to determine the influence on the consumer's spending
behaviour.

As opposed to a physical measurement, the object of a non-physical
measurement is often conscious of being measured. The mere fact that we
measure *influences* the subject. The subject will behave differently in a laboratory than in daily life. An example is given in Fig. 1.3 for an often used
(and abused) test animal: the rat. A measurement can even cause *irreversible
processes* in the object. Certain psychological experiments, for example, can

cause catatonic states in certain patients, resulting in mental disorders or even chronic schizophrenia. The effects mentioned above will generally make it difficult to reproduce measurements in the fields of human and social studies accurately, assuming that these measurements can actually be made reasonably selectively and objectively.

So, all in all, we see that when trying to measure non-physical quantities such as friendliness, intelligence, religiosity, tiredness, etc., one experiences various handicaps: The complexity of the system of which the measurement object is a part is large. The complexity of the object of measurement itself is high. Unknown dependencies and interactions exist between object and environment. No isolation experiments are possible to null these dependencies. No free manipulation of the variables affecting the measurement object is possible. It may not be possible to repeat the experiment. The choice of experiment is restricted for ethical, political or economic reasons. The subject may be conscious of being measured and alter its behaviour. Experiments may cause irreversible damage and may therefore be unacceptable.

Figure 1.3. Conducting unrestricted measurements on living creatures (laboratory rats) will meet with ethical objections. Here a brain electrode is used to stimulate the 'pleasure centre' in the hypothalamus. The response of the heart rate to electrical and mechanical stimulation of the brain is measured. This process is intended to be used as a 'reward' in learning processes. The validity of these measurements for the learning behaviour of a rat under normal conditions must be doubted.

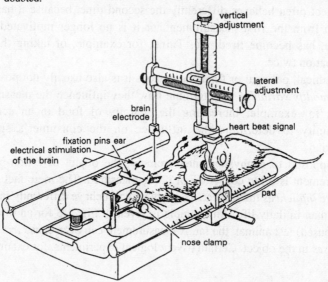

All these factors mean that, when developing measurements for non-physical sciences, one is very much limited in the procedures, techniques, experiments one can use compared to the physical sciences. This is the reason why no adequate measurements have been developed in most non-physical areas of human endeavour.

2

Measurement of physical quantities

In the following chapters we will confine ourselves to the study of measurement of physical quantities, such as mechanical, magnetic or electrical quantities. Measurement is then defined as a set of operations performed on a physical object or system (measurement object) according to an established, documented procedure (measurement method, strategy) using technical means (measurement system) for the purpose of determining some physical properties of the object or system. We can then define measurement as the acquisition of information (structural information as well as metric) about physical states or phenomena, by technical means.

The information that mankind has acquired through the ages is brought together in our 'concept' of the world. This world concept or image comprises the form of hypotheses, theorems and natural laws. Measurement therefore constitutes the source of our scientific knowledge. In other words: 'In physics, only that what can be measured exists' (Max Planck).

The information obtained by measurement can be present in the object of our measurement in two forms, passive or active. Information contained in the arrangement of matter is called *passive information*, for example information in a photograph or punch card, but also the information in the resistance of a resistor. *Active information* on the other hand is information present in the form of an energetic phenomenon. These informational energetic phenomena are called *signals*. Examples are electrical, optical and acoustic phenomena used for transferring information.

In the following we will deal with active and passive measurement objects in the most general terms possible. We cannot access passive information until it has been transformed into active information. This is done with the aid of some kind of energy source (illuminating a photograph with light, sending electrical energy into a resistor to measure it). The passive information present in the object of our measurement is then projected onto the energetic phenomenon; passive information becomes active information (the light reflected off the photograph, the current through or the voltage across the resistor). The active information can now be observed by our technical measurement means: the *measurement system*. Just as is the case with us humans, technical systems also exchange information (communicate) in the form of active information.

Measurements on 'passive' objects (i.e. where the measurement information to be obtained from the object is present in the passive form) requires an external energy source to *excite* or activate the relevant passive properties of the measurement object. This *exciter* supplies a stimulus to the measurement object, which in turn produces a response. If we know the stimulus, we only need to measure the response to characterise the measurement object. The response actually contains active information about both the object being measured and the stimulus. If the stimulus is unknown, both the response and the stimulus must be measured to obtain just that part of the passive information that was activated by the stimulus (see Fig. 2.1(a)).

Let us assume that the technical means we use for measuring, the measurement system, can measure the ratio of two physical quantities provided that they have the same physical dimension. However, the stimulus and response do not necessarily always have the same physical dimension. We therefore need a *reference* that is characterised by an accurately known relation between stimulus and response. This reference enables us now to measure a 'passive' object as shown in Fig. 2.1(a). An 'active' object (i.e. the information that is to be obtained from the

Figure 2.1. The acquisition of information from a measurement object.
(a) 'Passive' measurement object. (b) 'Active' measurement object.

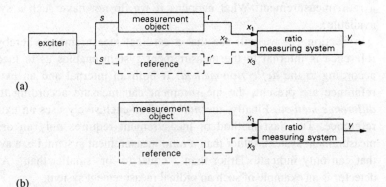

(a)

(b)

measurement object is active) does not, of course, require an exciter (see Fig. 2.1(b)), but again, we do need a reference in order to measure the object by means of a ratio measurement. This reference must produce a well-known signal of the same dimension as the signal to be measured. Examples of 'active' references are: accurate signal synthesisers, reference voltage sources and so forth. Examples of 'passive' references are: instrument transducers, attenuators and calibration resistors, etc.

It was shown in the previous chapter that a ratio measurement yields a linear relation $y = f(x)$ between the measurement result y and the magnitude of the input signal x, without offsetting the origin, thus $y = ax$ in which a is an unknown rational number. If, in Fig. 2.1, we measure x_1 and x_2 or x_1 and x_3, we can determine the ratios x_1/x_2 and x_1/x_3 respectively. These ratios are independent of a (a property of the ratio measuring system), provided that a is the same for both measurements. A ratio measurement is sufficient if one wants to know only the quotient of two quantities of equal dimension. However, if we wish to make a cardinal measurement we must have references at our disposal. These can be external references but they can also be incorporated in the measurement system.

Example: A moving coil indicating instrument producing a pointer deflection y which increases linearly with the input current x (assuming $y = 0$ when $x = 0$) is a ratio measurement instrument. If a cardinal measurement is to be performed with this instrument, a reference current source is required, whose output is accurately known. Another possible way of making a cardinal measurement of a current is by using an ammeter with a previously calibrated scale. This scale then indicates the actual magnitude of the current. This meter now has an internal reference, determined by the current-deflection sensitivity of the meter. In other words, the constant a of the ratio measurement is known.

In Fig. 2.1 we have given an example of a measurement system capable of a ratio measurement. What happens if we do not have such a system available?

In Section 2.2 we will see that a cardinal measurement whereby the reference is internal to the measurement system enables us to measure according to the *deflection method*. If both an internal and an external reference are present, the measurement can measure according to the *difference method*. Finally, the *null method* exclusively uses an external reference. This last method of measurement requires only an ordinal measurement system, rather than a ratio measurement system, i.e. a system that can only indicate 'larger than', 'equal to' or 'smaller than'. A null detector is an example of such an ordinal measurement system.

If we confine ourselves to quantitative physical measurement for which a ratio measurement system can be realised, the measurement essentially determines the relative size of one physical quantity with respect to another, of equal physical dimension. One quantity is called the 'measurand' and the other the 'the measure'. It is simplest to set the magnitude of the measure equal to unity. The magnitude of the measure for mass (the kilogram), for example, is also the unit of mass. Therefore, for a cardinal measurement a set of unit measures (units) is necessary, one for each different physical quantity. To facilitate international exchange and comparison of measurement results, these units should be accepted and utilised by all nations. We will return to this subject in more detail in Section 2.1.

Human observers need technical (measurement) means to make certain physical quantities observable (e.g. magnetic quantities) and to improve their perception (larger sensitivity, more objective). These means are designed to convert what is to be measured, without loss of information, into a quantity that cannot be misinterpreted by our senses. Usually only our sight and our hearing are regarded as 'addressable' senses. These technical means for measurement are generally referred to as 'measurement systems'. The term measurement system covers both the total measurement set-up as well as a single appliance or instrument. The field of designing measurement systems is called *instrumentation*. The purpose of this field is to realise measurement systems that utilise the available technological means and measurement methods in an intelligent manner. This requires knowledge of the measurement methods and principles that we will deal with in Section 2.2.

The result of a measurement must be a true representation of the magnitude of the measurand. Shortcomings in this representation in the form of errors will always occur though, consequently the result will not provide an infinitely accurate image of what is being measured. In Section 2.3 we will therefore deal with various sources of measurement errors.

2.1 Units, systems of units and standards

A ratio measurement enables us to relate the magnitude of an unknown physical quantity to that of a known physical quantity of the same dimension. The known magnitude of the quantity to which we refer the measurement is called the *measure*. For cardinal measurements the measure is internationally standardised and for simplicity set equal to unity. Therefore, in the case of cardinal measurements the measure constitutes the *unit* of the quantity which is being measured. The result of a cardinal measurement of a physical scalar quantity can therefore be written as:

$$magnitude = \{value\} \times [unit]$$

$$x = \{x\}[x]$$

We can now denote a vectorial physical quantity **x** as:

$$\mathbf{x} = \{x\}[x]\mathbf{e}$$

Here **e** is the unit vector (in the same direction as **x**) which contains the (dimensionless) information of the direction of x with respect to an (arbitrary) chosen coordinate system.

Between physical quantities of different dimensions there are mutual links in the form of definitions and natural laws. These mathematical relations determine the connection between the various physical quantities. This relation between different quantities can, for instance, take the form:

$$x = fA^aB^bC^c...$$

in which the numerical factor f is generally not equal to 1. The units $[A]$, $[B]$ and $[C]$ here define a new unit $[x]$. The quantity equation above can be split into a *numerical equation*,

$$\{x\} = f\{A\}^a\{B\}^b\{C\}^c...$$

containing only the magnitudes of the various physical quantities, and an *unit equation*,

$$[x] = [A]^a[B]^b[C]^c...$$

containing only the units of the quantities. For example, for the quantity equation $E = \frac{1}{2}mv^2$ with $m = 1$ kg and $v = 10$ m/s, the numerical equation results in $\{E\} = \frac{1}{2}\{m\}\{v\}^2 = \frac{1}{2} \times 1 \times (10)^2 = 50$. The unit equation becomes $[E] = [m][v]^2 = [m][l/t]^2 = $ kg m^2/s^2.

If the units in the unit equation are chosen such that no numerical factors other than 1 occur, as in the example given above, the units are called *coherent* with regard to the quantity equation. It is possible to devise a system of units that is totally coherent. A disadvantage of such a system is that some derived units become inconveniently large, compared to practical values. An example of this is the unit of capacitance in the Système International d'Unités, the farad. The largest values of technically realisable capacitors are in the order of 10^{-3} farad. For this reason, decimal prefixes such as mega, kilo, milli, micro, etc. are often used (see Section 5.1).

If k is the number of independent physical quantity equations that describe a particular area of physics (e.g. thermodynamics, mechanics, or electromagnetism) and n is the number of different quantities in the k equations, then $n - k$ quantities can be chosen freely as *base quantities* in a system of units suitable for that area of physics. The other quantities are *derived quantities* which follow from the base quantities and the k equations. The base quantities are related by definition to corresponding physical standards (see Section 5.1).

Determining which quantity in a system of units is a base quantity (with an associated base unit) is governed by the ease with which it can be measured, the logical structure of the system of units and the simple realisation of physical standards. The number of base quantities $n - k$ depends on the number of quantities n and the number of physical equations k that are considered to be independent. For example, in mechanics it is possible to choose F as a base quantity, as well as l, m and t. This would result in the equation which we are used to writing as $F = m\,a$, now becoming $F = cma$. Here c is a constant with a physical dimension $[c] = [F][t]^2/[m][l]$.

It is also possible to choose fewer base quantities, for example, only l and t, instead of l, m and t as is customary. The dimensions of mass and force are then fixed by the equations $F = c_1ma$ and $F = c_2m_1m_2/r^2$, with c_1 and c_2 numerical constants. This leads to a dimension of mass: $[m] = [l]^3/[t]^2$ and a dimension of force: $[F] = [l]^4/[t]^4$.

We can see that choosing 'too many' base quantities results in constants in the unit equations that have a physical dimension. The choice of the minimum number of base quantities gives numerical constants and, if the system is coherent, these constants become unity.

The foregoing discussion shows that a *derived quantity* is made up of products of powers of base quantities. The corresponding unit equation indicates which base units are used to arrive at the derived unit, or in other words, which physical *dimension* the respective unit has. The dimension $[A]$ of an area A, for example, is equal to $[l]^2$. If the base quantities of a system are l, m, t and I, the dimension $[V]$ of an electrical potential V will be $[m][l]^2/[I][t]^3$. Hence, analysing the dimensions in an equation gives a means of verifying the correctness of that equation. A correct equation of physical quantities has to satisfy the following conditions: the dimension of the left and right sides of the equation must be the same, the terms in an addition or a subtraction must have the same dimension and exponents and arguments of mathematical functions must be dimensionless. Take, for instance, the equation:

$$V_{avg} = \frac{1}{T} \int_{t}^{t+T} v(t) \, dt$$

for the time average (over a period T) of the instantaneous voltage $v(t)$. This equation gives rise to the unit equation: $[V_{avg}] = [t]^{-1}[v(t)][t]$. The equation $v = L \, di/dt$ yields $[v] = [L][I][t]^{-1}$. The exponent in the exponential function describing a semiconductor diode characteristic $I = I_s \{\exp (qV/kT) - 1\}$ must be dimensionless. This is correct, since $k = 1.3805 \times 10^{-23}$ J/K (see Appendix A.5.1). In the equation $v(t) = \hat{v} \sin(\omega t + \phi)$, ϕ must be dimensionless, just as the term ωt, which means that $[\omega] = [t]^{-1}$.

Note, however, that even when the dimensions in an equation describing the relation between different physical quantities are correct, it does not necessarily follow that the equation itself is correct; a mistake could still have been made in arriving at the equation. Correct dimensions are a necessity, but are not sufficient to guarantee a correct physical equation.

Although there are many possibilities for choosing a system of units, there are obviously numerous advantages in using an internationally accepted, standardised system of units. Such a system is the Système International d'Unités, or SI (see Section 5.1). The SI has seven base units. These are the metre, the kilogram and the second as units of respectively length, mass and time, the ampere as the unit of electrical current, the kelvin as the unit for thermodynamic temperature, the candela as the unit of luminous intensity and lastly the mole as the unit of the amount of a substance. In addition to these, two dimensionless numbers have been defined, i.e. the radian for angular measurement in two-dimensional space and the steradian for solid angular measurement. All other units are derived units. The SI is a coherent system of units. It is also a *rationalised system*, which means that the factors of 2π and 4π only appear in expressions that imply rotational symmetry or spherical symmetry respectively. The candela was introduced to simplify the measurement of the subjective intensity of light. This involves the (standardised) sensitivity of the human eye. Luminous intensity, however, can also be characterised by the wavelength and the energy of the light. So, in fact, the candela is a superfluous unit. The candela and the units derived from it form a separate section within the SI and are used solely in photometry.

In the past several systems of units were used side by side, each one specific to a certain field of physics. A widespread system was the cgs system (centimetre, gram, second). Correspondingly, there were two systems used in electrical engineering: the e.s.u. (electrostatic units) and the e.m.u. (electromagnetic units). In the e.s.u. the absolute permittivity of a vacuum ε_0 had no dimension and was equal to 1. The absolute permeability μ_0 followed from the relation $\varepsilon_0\mu_0c^2 = 1$, where c is the speed of light in vacuum. Capacitance therefore had the dimension of length. In the e.m.u. μ_0 was dimensionless and equal to 1. The units gauss and oersted which are still in use for magnetic induction and magnetic field intensity respectively, date from the same time. In the SI the absolute permeability is set to $4\pi \times 10^{-7}$ H/m, which fixes ε_0 at $\varepsilon_0 = 1/\mu_0c^2$ therefore $\varepsilon_0 \approx 8,854 \times 10^{-12}$ F/m (see Section A.5.1).

The terms *unit* and *physical quantity* are both abstract concepts. In order to use a unit as a measure, there must be a realisation of the unit available; a *physical standard*. A standard can be a tangible representation of the physical quantity as, for example, is the case for the standard measure of mass: the kilogram. A standard can also be defined by a standardised procedure of measurement using standardised measurement methods and equipment. This is the case for the standard measure of electrical current, which is based on measurements with a current balance. A third possibility is to use a natural phenomenon as a physical standard. The standards for length, time and electrical potential, for instance, are based on atomic processes (see also Section 5.1).

A primary standard is a standard that can be considered the highest metrological authority of a certain physical quantity. Each country has several of these primary standards. These national, primary standards are generally realised, maintained, preserved and improved in a national laboratory for weights and measures.

Ideally, these primary standards are left unperturbed and measurements are based on secondary or even tertiary standards (working standards). These are calibrated to higher (primary or secondary) standards. An even lower order standard is present in every instrument that can perform a cardinal measurement. Therefore, such instruments must be calibrated regularly, since ageing, drift, wear, etc., will cause the internal standard (also called the reference) to become less accurate. Accuracy is defined here as an expression of the closeness of the value of the reference to the primary standard value.

In the following sections several standards of electrical quantities will be discussed.

Electrical potential difference

The primary standard for electrical potential difference used to be provided by an electro-chemical *standard cell* (the Weston standard cell), see Fig. 2.2. The voltage of a Weston cell is approximately 1.01860 V at 20 °C, with an inaccuracy of 3×10^{-6} under optimal conditions. Optimal conditions mean a variation of temperature less than 10^{-3} K, no load, no vibrations or jolting. The cell must remain in an upright position. A Weston standard cell has a very long thermal after-effect. After the cell has been heated to 30 °C it can take 6 months before it is totally stable within 0.3 µV of the original value at 20 °C. Furthermore, a Weston cell will age, resulting in an increase in the internal resistance ($R_i \approx 500$–$1000 \ \Omega$), and a small decrease of a few µV in the output voltage (during the first years).

For lower standards, zener references will suffice. A zener reference is biased at a constant current and kept at a constant temperature. It can produce an inaccuracy of less than 10^{-5}. Zener references are particularly useful as transfer standards. They are often used in portable instruments.

The vulnerability and relatively large inaccuracy of standard cells led to a search for a voltage standard that is more constant and more reproducible. This resulted in the use of the *Josephson effect* for voltage standards. This effect, predicted in 1962 by the British student Brian Josephson, occurs when two superconductors are brought close enough together (ca 10^{-9} m) for the quantum wave functions to become weakly coupled.

Usually thin films of lead, which are cooled below the transition temperature, are used as superconductors. These films are separated by a 1 nm thick oxide layer. It is possible for electrons to tunnel across the resulting junction. This electrical tunnel current can be influenced by a high-frequency electromagnetic field (frequency $= f_0$) perpendicular to the junction. This gives a staircase characteristic for the current as a function of

Figure 2.2. Schematic illustration of a Weston standard cell, historically used as the primary standard for electrical potential difference.

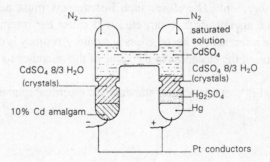

the voltage, as shown in Fig. 2.3. This function is characterised by abrupt increases at precisely quantised Josephson voltages V_J. The voltage $V_J(n)$ corresponding to the nth step (with n a whole number) depends on the frequency f_0 according to $V_J(n) = nf_0(h/2q)$, where h and q are fundamental physical constants, h is Planck's constant: and q is the charge of an electron. The value of $2q/h$ for standardisation purposes has been fixed at $2q/h = 483\,597.9$ GHz/V. By connecting junctions in series, a voltage of approximately 10 mV can be obtained, with an uncertainty of only 4×10^{-7}. ($f_0 \approx 10$ GHz at a temperature of 4 K.)

Electrical current

Electrical current is standardised by measurement with an instrument called a 'current balance'. This device measures the electro-magnetic force between two current carrying coils (one fixed, one moving) by balancing it with the force of gravity, acting on a known mass (see Fig. 2.4). The force between the coils is given by $F = I^2 \, dM/dx$, in which M is the known mutual

Figure 2.3. The current characteristic of a Josephson junction as a function of the junction voltage, currently used as a primary standard of electrical potential difference (voltage).

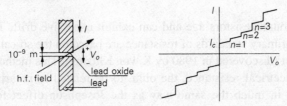

Figure 2.4. (a) Schematic representation of a current balance. (b) Detailed view of the coil arrangement in a uniform field generated between the upper and lower fixed coils.

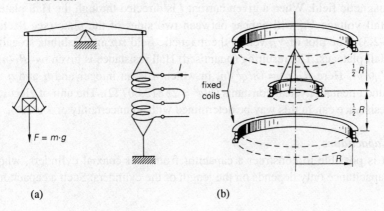

(a)   (b)

induction of the coils and x the known distance between them. The differential quotient dM/dx is determined by the known geometry of the coils. The inaccuracy that can be achieved with a current balance is approximately 3×10^{-6}. The current balance is an example of a 'passive' reference (passive in the sense of information contained in the information of the reference) as explained in Fig. 2.1.

Electrical resistance

Standards of electrical resistance are resistors wound of special alloy wire giving a minimal temperature effect. An example of such an alloy is evanohm, consisting of 74% nickel, 20% chromium and 6% aluminium and iron. This alloy is frequently used for high resistance standards (10 kΩ). For low resistance values (1 Ω) manganine (86% Cu, 12% Mn, 2% Ni) or constantan (54% Cu, 45% Ni, 1% Mn) is often used. Higher-order resistance standards are kept at very accurately stabilised temperatures by thermostats. Resistance standards are sensitive to moisture and heating through dissipation within the standard. Low-value standards are equipped with two pairs of connectors: current connectors and voltage connectors. This is done to avoid the effects of contact and lead resistance when measuring the resistance. An uncertainty of approximately 1×10^{-6} can be achieved.

Wire-wound resistors age and can exhibit excessive drift. Therefore, at present, primary standards of resistance are based on the so-called quantum Hall effect (discovered in 1980 by K. von Klitzing). This method allows the unit of electrical resistance, the ohm, to be related to invariant physical constants in much the same way as the Josephson effect for electrical potential difference.

The quantum Hall effect occurs when a semiconductor Hall plate, with a high charge carrier mobility, is cooled to 1 K and placed in a strong magnetic field. When a given current I is directed through the Hall plate, a Hall voltage V_H will appear between two sides of the plate (see Section 3.2.3). The plot of V_H versus the magnetic field strength exhibits so-called Hall plateaus. The resulting (quantised) Hall resistance is given by $R_H(n) = V_H(n)/I$. Here $R_H(n) = (h/q^2)/n$, in which n is an integer and h and q are fundamental physical constants ($h/q^2 = 25\,812.807\ \Omega$). The unit of electrical resistance can in this way be determined with an uncertainty of 2×10^{-7}.

Capacitance

It is possible to construct a capacitor from four coaxial cylinders, whose capacitance only depends on the length of the cylinders. Such a capacitor is particularly suitable as a standard of capacitance, since only the length has to be determined accurately. With the aid of optical interferometry this can be done extremely accurately. These so-called Thompson–Lampard cylinder capacitors can achieve an inaccuracy of less than 10^{-8}. A disadvantage, however, is the fact that the capacitance is small (approximately 1.9 pF per metre). For lower-order standards other configurations of electrodes are used, which provide larger capacitance values (10–1000 pF), but also come with larger uncertainties.

Inductance

Accurate standards of inductance are difficult to realise. This is caused by the many parameters that determine the relatively complex geometry of a coil, all of which influence the accuracy of the inductor. Furthermore, power losses occur, due to wire resistance, proximity effects and eddy currents, which add to the inaccuracy. Currently available standards of inductance have an inaccuracy of about 10^{-5}.

Frequency

The standard of frequency is based on the quantum mechanical effect that electrons in an atom can only occupy one of a limited number of discrete energy levels. If an electron jumps to a higher or a lower energy level, the difference in energy ΔE of the photon that is absorbed or emitted, respectively, is related to the frequency of the photon by the expression $\Delta E = hf_0$. When atoms are irradiated with electromagnetic energy of frequency f_0, many electrons will pass to higher energy levels. Caesium-133 has a suitable transition between two energy levels of the ground state (from $l = 3$, $m = 0$ to $l = 4$, $m = 0$, in which l is the orbital angular momentum quantum number and m is the azimuthal quantum number). The frequency associated with this transition is, by definition, $f_0 = 9.19263177160$ GHz. The unit of time is defined as the interval in which f_0 cycles fit exactly. The atoms that have made the required transition are selected by deflection in a magnetic field. The (neutral) atoms are then ionised by a filament. The ions cause a current in a detector that is a measure of the number of ions per second. A crystal oscillator in a feedback loop is used to adjust the frequency of the standard to precisely that frequency at which most transitions occur. The quality factor Q of the so-tuned standard is approximately 2×10^7. The relative uncertainty of this atomic frequency standard can be as small as 10^{-12}.

2.2 Measurement methods

In order to conduct measurements in an optimal fashion, it is essential to be familiar with the most important methods, principles and strategies of measurement. A measurement is performed optimally if the result and the desired accuracy are obtained with the most simple means and strategies. Often, a different method of measurement will permit the use of simpler and therefore less expensive equipment. There are more measurement methods than examined in this section; we will only discuss the ones most commonly used. It is not always possible to define clearly the divisions between the various methods; often a measurement employs several methods simultaneously.

The first three methods which we will describe here are related to the extent to which the measurement result is influenced by the reading or display of the measurement instrument. In order of decreasing influence, these three methods are: the deflection method, the difference method and the null method.

Deflection method, difference method and null method
With the deflection method the read-out of the measurement device used entirely determines the result of the measurement. The difference method measures (indicates) only the *difference* between the unknown quantity and the known, reference quantity. Here, the result of the measurement is partially determined by the read-out of the measurement device used and partially by the reference quantity. Finally, with the null method the result is entirely determined by a known reference quantity. The read-out of the measurement instrument is used only to adjust the reference quantity to exactly the same value as the unknown quantity. The indication is then zero and the instrument is therefore used as a null detector or zero indicator.

We have seen in the first part of this section that a measurement system which can be used for making cardinal measurements must have an internal reference. The contribution of the displaying instrument to the total error in the measurement's end result decreases with each described method. The internal reference of the overall measurement system can therefore be less and less accurate, at the expense of an increasing accuracy required of the 'external' reference. NB: The measurement system, i.e. all means used to make the measurement, is thought of here as comprising the displaying instrument as well as the required reference external to this instrument.

Examples: The length of a rod of approximately 101 mm must be measured with a relative inaccuracy of $\pm 10^{-4}$. If the deflection method were to be used, a measuring instrument such as vernier callipers with a relative inaccuracy of $\pm 10^{-4}$ must be used. Since the accuracy of existing callipers is inadequate, the difference method is applied, as illustrated in Fig. 2.5(a). The available reference length in the form of a gauge block, may be, for instance, 100.000 mm $\pm 10^{-5}$. The difference in length of approximately 1 mm is measured with a dial gauge. This difference must be determined with an accuracy of $\pm$ 10 μm. The relative inaccuracy of the dial gauge may therefore be $\pm 10^{-2}$. This a factor 100 larger than the required relative inaccuracy for the measurement with the deflection method.

The null method could be used if we have accurate reference lengths which when stacked are as long as the rod being measured. The dial gauge can be used as a null indicator here, for which the inaccuracy of the reading is irrelevant. Provided that a small difference in length still causes an obvious deflection of the dial gauge, the rod can still be measured with a relative inaccuracy of 10^{-4}.

A second example is indicated in Fig. 2.5(b). An unknown pressure (measured) is compared to a known pressure which is generated by a predetermined weight on a plunger (with a known cross sectional area and weight). Any displacement of the membrane between the two halves of the measuring chamber is measured by a differential capacitor (see Section 3.2.1). The weight on the plunger is adjusted until the capacitances of the two halves of the capacitor is equal (null method). The measured pressure is equal to the total weight divided by the area of the plunger. Note that special precautions must be taken to reduce the friction between the plunger and the chamber wall.

One can easily imagine that, if the pressure chamber is calibrated with a known pressure difference, it will be possible to measure pressure differences without totally compensating the measured pressure. Thereto, the displacement of the elastic membrane is characterised by the capacitance difference ΔC_{out} versus pressure difference ΔP. The resulting ΔP is added to

Figure 2.5. Difference method used to measure (a) length, (b) pressure.

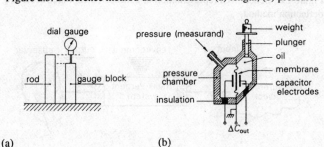

(a)   (b)

the pressure determined from the external weight and the cross sectional area of the plunger.

The *deflection method* normally provides simple measurements and quick results. In Fig. 2.6 a generalised representation of the deflection method is given. An essential aspect is that there are four subsystems involved: the *measurement object*, the *measurement system*, the *observer*, and the *environment* of the measurement system. These subsystems exhibit interactions, some of which are desirable, others are undesirable. The desirable interactions are those that transfer the useful information: the measured object influences the measurement system and the measurement system in turn influences the observer. Unfortunately, these interactions are always *mutual*. The measurement system will influence the measured object (for instance, by loading) and the observer can influence the measurement system. For instance, the observer does not read a dial with only one eye, held in a plane through the pointer, perpendicular to the dial in order to avoid parallax. The surroundings also influence the measurement system (take, for instance, the effects of temperature, moisture and pressure on the measurement system).

The achievable measurement accuracy of the deflection method is usually not very high. The achieved accuracy is limited by the influence of the measurement system on the quantity x which is being measured (matching error), by the accuracy of the transfer characteristics of the measurement system (system accuracy), by the accuracy with which the result y can be read (reading accuracy), and by the interfering influences of the environment on the measurement system (influence errors).

The *difference method* requires a less accurate indicating instrument, at the expense of requiring a reference quantity. The value of this reference quantity must be known accurately.

The measurement consists of a principal part p (the value of the reference quantity) and a remainder r, which is the reading of the indicating

Figure 2.6. Generalised representation of a measurement made according to the deflection method.

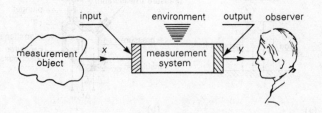

instrument. If Δp is the absolute error of the principal part p, and Δr the absolute error of the remainder, we call the measurement optimal when $\Delta p = \Delta r$. This means that if the remainder is the nth fraction of the principle parts: $r = p/n$, the relative error $\Delta r/r$ in the measurement of the remainder may be n times larger than that in the principal part. It stands to reason that the difference method is uniquely applicable to cases for which it is more important to determine the *difference* in value, rather than the *absolute* value. The difference method is also often used for accurately measuring small variations superimposed on a large constant value, for example, when measuring the long-term stability of the output voltage of an electrical power supply .

The *null method* requires an instrument with sufficient sensitivity and a stable 'null' indication (the null may not drift *during* a measurement). Usually, the accuracy of null method measurement is determined by the reference, since practical null detectors can achieve a very high degree of resolution and stability.

It is not necessary for the deflection y of a null detector to be linearly dependent on the input signal x. Often, the transfer characteristic of a null detector is highly non-linear, to allow a larger range of measurement (sensitive around scale zero; increasingly insensitive for larger inputs). The important aspect of a null detector is the sensitivity of the y–x characteristic around the origin. In order to minimise the influence of noise, hum, harmonics and other forms of distortion in the input, the null detector generally has a small bandwidth (which unfortunately causes the detector to be slow). Such a frequency-selective null detector mostly has a tunable band-pass input filter.

Interchange method and substitution method
In both the interchange method and the substitution method two measurements are performed in succession. With the *interchange method* the unknown quantity and the known quantity are both used simultaneously in each measurement. With the substitution method the known and unknown quantities are used separately and successively.

The *substitution method* first measures the unknown value of the measurand, which results in a certain indication deflection or reading of the measurement system. Then, the unknown quantity is replaced by a known and adjustable quantity, which is adjusted such that exactly the same measurement result is found. The indication of the measurement system is used here as an intermediate only. The characteristics of the measurement system should therefore not influence the measurement. Only the time stability and the resolution of the system are important. The resolution determines the 'fineness' with which the unknown and known quantities can be made equal, the short-term stability determines the 'drift' of the measurement system between the two measurements.

Calibration of a measurement system is, in fact, an application of the substitution method. First, the system is calibrated with a known quantity. An unknown quantity can then be measured accurately if its magnitude coincides with the calibrated points. A substitution method is also often used as a simple means for establishing equality, where the accuracy of the measurement system used to determine 'equal' does not matter.

The *interchange method* is a method of measurement in which two, almost equal quantities in a measurement system are exchanged in a second measurement. This method can determine both the magnitude of the difference between the two quantities and the magnitude of a possible asymmetry in the measurement system. An example of this method is checking the equality of the two arms of a balance by swapping the weights on either side.

Compensation method and bridge method

On the basis of the number of power sources involved in a measurement, two other methods can be distinguished: the compensation method and the bridge method.

The *compensation method* is a method of measurement that removes the effect of the unknown quantity on the measurement system by compensating it with the effect of a known quantity. This is done in such a way that the unknown quantity is no longer influenced by the measurement system when full compensation is reached. If the unknown effect is compensated completely, no power is supplied to or withdrawn from the unknown quantity; the unknown quantity is not loaded by the measurement. The degree of compensation can be determined with a null indicator. The compensation method therefore requires an auxiliary power source that can supply precisely the power that otherwise would have been withdrawn from the measured quantity. Thus, for the compensation method, we need two power sources.

Examples. To compensate an electrical potential difference V_x in Fig. 2.7(a), a voltage αV_N is shunted from a supply voltage V_N by a variable potentiometer. A null indicator is used to adjust the potentiometer so that αV_N becomes equal to V_x. When $\alpha V_N = V_x$ the compensation is complete and V_x is no longer loaded. The voltage source V_N is the auxiliary power source. In order to be able to measure V_x accurately, V_N and α must be accurately known.

In Fig. 2.7(b) the compensation method is applied to the measurement of an unknown force F_x. Before full compensation is reached, the force F_x pushes the extension rod down by an amount Δx, thus supplying energy $(F_x \Delta x)$ to the system. This causes a valve to close a little further and the pressure beneath the pressure plate A to increase and, with it, the compensating upward force on the rod. Equilibrium is reached when the pressure under A compensates the force F_x exactly. Now the energy is supplied by the air pump (air supply P_N). The compensation therefore occurs automatically here.

Note that the illustrated method of measuring the pressure αP_N with a manometer is, in fact, an application of the deflection method!

The use of two separate power sources, such as V_x and V_N in Fig. 2.7(a) is essential for the compensation method. A disadvantage of this method is however, that if one of the power sources tends to drift, there will no longer be exact compensation.

Compensation can also be used to eliminate disturbances. In a differential electronic amplifier with dual bipolar transistors (a long-tailed pair) compensation reduces the effect of temperature variations in the base-emitter voltage of approximately 650 mV with a temperature coefficient of − 2.5 mV/°C to approximately 1 mV with a temperature coefficient of ± 2 μV/°C.

NB The difference method and the null method also make use of the compensation method. In the difference method, the compensation is only partial but in the null method it is complete.

The *bridge method* makes use of a bridge circuit. This is an impedance network with two ports, characterised by the fact that the transfer between the two ports is zero for certain values of the impedances (without any of the impedances being equal to zero). The condition which the impedances

Figure 2.7. Use of the compensation method: (a) electric, (b) mechanical.

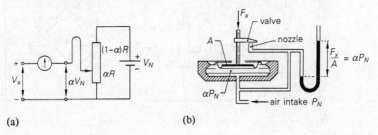

(a)   (b)

satisfy when the transfer equals zero is called the *null condition* or the *balanced condition* . In order to adjust the circuit to the zero situation, a power supply is connected to one of the ports, and a null indicator to the other. It can be shown that the null condition does not depend on the power delivered by the power supply, the circuit's internal impedance or the internal impedance of the null detector. Fig. 2.8 illustrates that the bridge method is, in fact, an application of the compensation method using only one power source.

The bridge circuit with four resistors was invented in 1833 by S.H. Christie. However, this configuration was named after Sir Charles Wheatstone, who first used it in 1843 for measuring resistance. Originally, the null indicator was called 'the bridge', as it connects either side of the circuit. Later though, the name was passed onto the impedance network used and this method of measurement.

Besides being used for measuring *electrical* impedances, the bridge method is also useful for measuring other impedances, such as thermal, hydraulic, acoustic, etc.

The term 'impedance' can be generalised to cover other systems besides electrical ones. Accordingly, the following holds for the impedance $\overline{Z}$ of

	mechanical systems:	translation	$\overline{Z} = \overline{F}/\overline{v}$
		rotation	$\overline{Z} = \overline{M}/\overline{\omega}$
	thermal systems:		$\overline{Z} = \overline{\Delta T}/\overline{I}_h$
	pneumatic, acoustic systems:		$\overline{Z} = \overline{\Delta p}/\overline{I}_v$
	electromagnetic systems:		$\overline{Z} = \overline{E}/\overline{H}$.

In these expressions $\overline{Z}$ is the generalised complex impedance and:

F = force	M = moment	ΔT = temperature difference
v = velocity	ω = angular velocity	I_h = heat flow
Δp = pressure difference		E = electrical field strength
I_v = volume flow		H = magnetic field strength.

Figure 2.8. From a voltage compensation network (a) to a Wheatstone bridge (b).

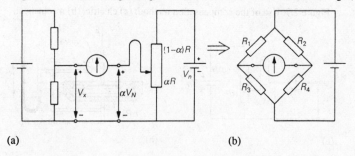

(a)   (b)

Analogy method

This method makes use of a model of the object from which we wish to obtain measurement information. Measurements made on the model then provide information about the unknown object, as long as the model corresponds with the object in certain essential points. This analogy method is most often used when measurements of the actual object are not possible, extremely difficult, time consuming or costly.

One class of models used is that of the *mathematical models*. The model is here described by the same mathematical equations as the actual object. For instance, a mechanical balance with arms of different lengths, can be considered as a model of an electrical bridge circuit of resistors (Wheatstone bridge). Let the lengths of the arms be l_1 and l_2, the masses on either side of the balance m_1 and m_2, and the gravitational acceleration g, then equilibrium is reached when $m_1 l_1 g = m_2 l_2 g$. This condition is independent of g, as g can be eliminated from the equation. The gravitational acceleration acts as the 'power supply' of the model and the pointer of the balance is the null detector.

Another class of models is that of the *scale models*, which are linearly scaled enlargements or reductions of the object to be measured. This type of model is often used, for instance, for investigating the acoustics of large halls, etc.

A third class consists of models which are *scaled non-linearly*. The enlargement or reduction is such that only certain figures of merit remain intact. Examples of these *figure of merit models* are irrigation and wind tunnel models.

An important application of the analogy method is the analogue computer, in which a physical problem is modelled by electronic circuits, which are governed by the same mathematical equations.

Since the measurement of a model or analogue is used quite frequently, we will discuss the way in which mathematical analogies of systems from different fields of physics are realised.

It is possible to convert similar systems from different areas of physics into one another in such a way that the mathematical description of their behaviour is identical. Such an analogous mathematical description can be found in the generalisation of electrical circuit theory. In the following, we will confine ourselves to discussing one analogue which is closely related to measurement theory, i.e. the analogue based on a describing physical system with the aid of V-quantities and I-quantities. A V-quantity is a (signal) quantity which is measured by connecting the measurement system *parallel* to the respective element, i.e. parallel to the port or the connectors, *across* which we wish to know the V-quantity. Another name for a V-quantity is an 'across quantity'. An I-quantity is measured by connecting the measurement system *in series* with the element *through* which the unknown quantity flows. Another name for an I-quantity is a 'through quantity'. Across quantities and through quantities in a particular domain of physics are chosen in such a way that the *product* of corresponding across and through quantities is equal to the *instantaneous power* transferred by these two quantities. If we assume that the across and through quantities are complex, just as in electrical circuit theory, then (for a linear system) the *quotient* of corresponding across and through quantities is the *impedance* of the element with $\overline{V}_h$ across the element and $\overline{I}$ through flowing it. Examples of across quantities are electrical potential V, velocity v, angular velocity ω, temperature difference ΔT, pressure difference Δp. Examples of through quantities are electrical current I, force F, moment M, heat flow I_h and volume flow I_v.

In thermal systems, generalised quantities must be applied cautiously. The product of $\overline{\Delta T}$ and $\overline{I_w}$ cannot be instantaneous power, as already the physical dimension of heat flow $[I_w]$ = watt. Furthermore, there is no such thing as a thermal inductor, and a thermal resistor does not dissipate energy. In Section 5.4 the across and through quantities and the corresponding impedances for several fields are listed.

An advantage of this particular method of generalisation of physical systems by means of across and through quantities is the fact that the circuit configuration of the various systems remains intact. Thus, the elements of these circuits can replace one another. For instance, an electrical capacitance C corresponds to a mass m in a mechanical translational system, or an inertia J in a mechanical rotational system (see Section 5.4). This analogy between physical systems is sometimes called the mass–capacitance analogy. A drawback of this analogy is the fact that the impedance of mechanical systems is the reciprocal of impedance as it is traditionally introduced in mechanical engineering (based on other analogies, see Table A.5).

Repetition method
With this method several measurements of the same unknown quantity are conducted, each according to a different procedure. Most fundamental physical constants have been measured in several different ways, to prevent the possibility of making the same (systematic) errors, specific to a certain type of measurement. Different (correctly applied) methods of measurement will produce similar results, but the measurement errors in the results will

be independent of each other. This will yield an indication of the reliability of the measurement.

Enumeration method

The enumeration method is a method for determining the ratio of the magnitudes of two quantities (the known and the unknown) by *counting*. Only numbers of objects, patterns or events can be counted. Physical quantities of a given physical dimension must be *measured*. Measurement involves making errors, counting does not (mistakes excluded). The enumeration method is used, for instance, for measuring frequency. The frequency of a periodical signal is measured by simply counting the number of periods which fall within an accurately determined reference time interval. The enumeration method is also important for analogue-to-digital conversion (see Section 3.3.6). Sometimes, though, it is easier to change from enumeration to measurement; for instance, when determining the total number of screws in a package not by counting them, but by weighing them.

Measurement strategies

In addition to the methods of measurement discussed above, we will now examine several widely used *measurement strategies*. It is, in fact, not always possible to measure the desired physical quantity directly. If, for example, a quantity fluctuates more quickly than the measurement system can follow, correct information cannot be obtained; the frequency spectrum of the measured signal is wider than the bandwidth of the measurement system. The reverse can also occur: the measurement system can have a considerably larger bandwidth than the signal's spectral width. In this case, the measurement system is not used optimally. In the first mentioned situation, *coherent sampling* will enable us to measure certain signals with a wide frequency spectrum, by means of measurement systems with a small bandwidth. In the second situation above, the *multiplexing* of a number of measurement signals will lead to a more optimal utilisation of the measurement system.

1 Coherent sampling

This measurement strategy enables us to process a measurement signal with a spectrum width F, which is considerably larger than the bandwidth B of the measurement system, provided that the signal is *periodic*. By taking samples of the actual measurement signal at intervals that are just slightly larger than n periods of the signal (n integer), the shape of the signal can be preserved and a true representation can be obtained. If we denote the interval between the samples as $nT + \delta$, in which T is the period of the actual signal, the period of the reconstructed (from the samples) signal will be $(nT + \delta)T/\delta$. This means that the frequency has been reduced by a factor $\delta/(nT + \delta)$. Fig. 2.9 illustrates this principle. The number n of skipped periods and the ratio T/δ (the number of samples in one period of the reconstructed signal) are chosen in such a way that the reconstructed signal, which is the envelope of the peaks of the samples, has a frequency spectrum which is smaller than the bandwidth of the measurement system used to process the sampled signal. This type of sampling is used in stroboscopic measurements and in sampling oscilloscopes. A sampling oscilloscope with a bandwidth of 20 kHz, for instance, can display (periodic) electrical signals with a frequency of up to 15 GHz (see Section 4.4).

2 Random sampling
Coherent sampling, as discussed above, requires some kind of provision that ensures that the samples (1, 2, 3, ... in Fig. 2.9) are taken at precisely the right instants. If we are only interested in *amplitude information*, however, and not in the shape of the signal, the samples can be taken at arbitrary moments: *random sampling*. In this manner it is very easy to determine, for example, the RMS-value of a signal with a wide frequency spectrum. The signal does not have to be periodic here. Another application is in obtaining the amplitude distribution function of a signal. More generally one may state that the statistical parameters of the signal amplitude are not affected

Figure 2.9. Coherent sampling of a signal shown in (a) with fundamental frequency $f_0 = 1/T$. The sampling interval is chosen to be $T + \delta$ (so $n = 1$). The lower trace (b) shows the sampled signal with a frequency $f_s = f_0\delta/(T + \delta)$. This signal can be processed by a bandwidth $B \ll f_0$ if $\delta \ll T$.

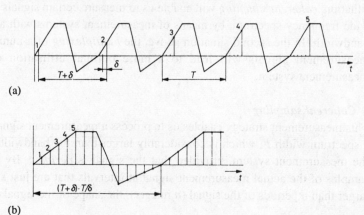

(a)

(b)

by random sampling (mean, standard deviation, curtosis, etc.). Sometimes 'random' sampling is realised by sampling at a certain frequency unrelated to the input signal. This can still produce correlation between the signal and sampling frequency and therefore errors may occur in the amplitude sampling. To minimise this, the sampling frequency is usually swept back and forth between the limits. For random sampling, the bandwidth B of the measurement system may again be much smaller than the frequency spectrum width F of the signal.

3 Multiplexing

This measurement strategy is a means of processing several signals, either simultaneously (frequency multiplexing) or sequentially (time multiplexing). This method can be used when the bandwidth B of the measurement system is much wider than the frequency spectrum width F of the measurements signals. Fig. 2.10 gives an illustration of *time multiplexing* as it may be used, for instance, in an oil refinery. The analogue input signals come from various measuring points in the refinery where, say, temperature, pressure and flow are measured. These signals vary so slowly that a fast measurement and control system can process many of these signals

Figure 2.10. Time multiplexing for process control.

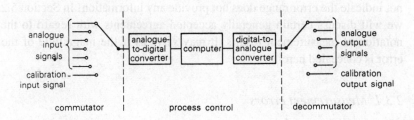

Figure 2.11. Frequency multiplexing ($F \ll B$).

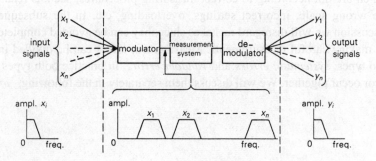

sequentially and can put the resulting individual outputs on as many lines as there are inputs (see Fig. 2.10). In Section 4.5.7 we will return to these kinds of measurement systems.

Frequency multiplexing shifts the frequency band of a narrow-band measurement signal, by modulation, to a different frequency band. This is done such that the spectra of several converted measurement signals will fit side by side in the frequency band, without overlap. A demodulator at the output of the measurement system is necessary to restore each signal to its original frequency band. An example of this method is depicted in Fig. 2.11. This method is often used in telemetry (transmitting measurement signals) and telephony.

2.3 Error theory

Unfortunately, every measurement is afflicted with measurement errors, i.e. there is always a difference between the measured value and the true value of the physical quantity under observation. The value of an error can never be known exactly; it can only be estimated. Measurements can only be correct to a certain degree and therefore the magnitude of the measurement error must always be given together with the measured value to constitute meaningful measurement information. A measurement result which does not indicate the error range does not provide any information! In Section 5.2 we will discuss certain generally accepted agreements with regard to the notation of measurement results. It may *seem* as if the magnitude of the error is concealed here!

2.3.1 Measurement errors

Discrepancies in a measurement can be divided into two categories: *mistakes* and *measurement errors*. Mistakes arise from operator actions which are not according to correct measuring procedures, such as reading the wrong scale, incorrect settings, overloading, etc. In the subsequent discussion we will disregard mistakes since they can be avoided completely by making careful measurements. Measurement errors can be divided into two types: *systematic errors* and *random errors*. In practice both types of error occur together. We will discuss them separately in the following.

Systematic errors

If a particular physical quantity is measured several times with a given measurement system under the same conditions, we will find that errors occur which are of the same magnitude every time. These errors are called systematic errors. Examples of systematic errors are loading errors and mismatch errors which cause the measurement system to influence the object under test. Another example is errors that are due to imprecise knowledge of the system's transfer characteristics; system errors.

Systematic errors can be traced by a careful analysis of the measurement path: from measurement object, via the measurement system to the observer. Another way to reveal such systematic errors is to perform the measurement according to a totally different method using different equipment (see Section 2.2, repetition method). Often, this kind of error can be minimised by carefully calibrating the entire measurement path. Also, if the underlying mechanism causing the error is known, a correction factor can be taken into account to reduce the effect of the error.

Fig. 2.12 shows an example of a systematic error due to loading the object under test. We wish to measure the temperature difference V_s (temperature difference is an across quantity). The internal thermal resistance of the object is R_s and that of the measurement system is R_i. However, the measurement system will indicate the temperature difference V_i at the input terminals of the measurement system; it does not know the actual object temperature V_s. Thus, $V_s = \alpha V_i$, in which $\alpha = (R_s + R_i)/R_i$. The measured temperature must therefore be corrected with a factor α.

In Fig. 2.12(b) R_i is equal to the thermal resistance of the measurement probe tip at the ambient temperature. If the end of the probe tip is not smooth, or not held flush onto the housing of the transistor, there will be an additional transfer resistance R_0 (in series with the input of the measurement system of Fig. 2.12(a)). R_i and R_s can be estimated from the dimensions and

Figure 2.12. Correction of a systematic loading or mismatch error. (a) Network analogy of a temperature measurement. (b) Temperature measurement of a power transformer.

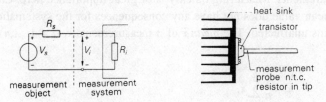

(a)   (b)

materials in question. For instance: measurement probe stainless steel, heat resistivity $r = 10 \times 10^{-2}$ K m/W, diameter 2 mm, length 3 cm ($R_i \approx 9.5 \times 10^2$ K/W); transistor: aluminium, $r = 4.9 \times 10^{-3}$ K m/W ($R_i \approx 1.2$ K/W).

Initially, immediately after the measurement probe is positioned on the transistor, the thermal capacitance of the probe has to be 'charged'. The indicated temperature is then lower than the final temperature. The case described above applies to the 'steady state', (theoretically infinitely) long after the probe is in position.

Random errors

Random errors are errors which vary unpredictably for every successive measurement of the same physical quantity, made with the same equipment under the same conditions. They are usually caused by a large number of factors, all of which influence the measurement independently. We cannot correct random errors, since we have no insight into their cause and since they result in random (non-predictable) variations of the measurement result. Examples are the errors an observer makes when reading analogue scales (such as a thermometer), adjustment or alignment errors when adjusting the zero setting or alignment of a measurement instrument (e.g. nulling a bridge) and round-off errors. When dealing with random errors we can only speak of the *probability* of an error of a given magnitude. Luckily enough *probability theory* (and statistics) enables us to make certain assertions in the presence of random errors. Both systematic and random errors can be considered to be caused by an interfering signal superimposed on the true signal which is to be measured. For random errors this interfering signal fluctuates haphazardly, whilst for systematic errors this signal has a constant but unknown value. The first signal has to be described in probabilistic terms; the latter in deterministic terms. Unfortunately, this deterministic character makes it more difficult to detect systematic errors.

It is possible to reduce the effect of random errors in the individual measurements by measuring several times and using the mean value of the measurements as the end result. Obviously this is only possible if the quantity which is to be measured does not vary during the course of these measurements. Measuring quickly is of great importance here. Calculating the mean value does not have any consequences for the systematic error; it remains unaffected. The *mean* $\bar{x}$ of n measurements x_i ($i = 1,...,n$) is given by:

$$\bar{x} = \frac{1}{n} \sum_{i=1}^{n} x_i$$

The mean $\bar{x}$ is the best possible estimate of the physical quantity with constant value x, based on n individual measurements or samples, x_i (with i = 1,...,n), if these are afflicted with random errors. This can be deduced from the fact that:

$$\sum_{i=1}^{n}(x_i - \bar{x}) = 0$$

Thus, the sum of all deviations $x_i - \bar{x}$ is zero. In addition,

$$\sum_{i=1}^{n}(x_i - \bar{x})^2$$

is minimal. So the deviations of the samples x_i with regard to the mean $\bar{x}$ or, in other words, the dispersion of $\bar{x}$ is minimal, as can be shown by simple substitution.

A measure for the dispersion of x around the mean $\bar{x}$ (which is a measure for the central tendency) is the *variance* σ_x^2 which is defined as:

$$\sigma_x^2 = \frac{1}{n-1}\sum_{i=1}^{n}(x_i - \bar{x})^2$$

The quantity usually given is the square root of the variance; this is called the *standard deviation* σ_x. We can display the measured samples x_i (i = 1,...,n) of a quantity x afflicted with random errors in a *bar graph*. We obtain a bar graph by dividing the range of x (x_{min}, x_{max}), which comprises all measured samples x_i, into smaller class intervals of width Δx, and then plotting the number of samples $N(x)$ falling into these smaller intervals (x, x + Δx), as a function of x (see Fig. 2.13). Normally the width of a class interval is chosen in accordance with:

Figure 2.13. Histograms: (a) correct class interval width; (b) interval width too small; (c) interval width too large.

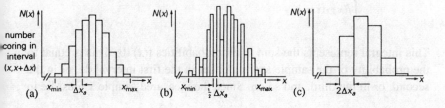

$$\Delta x = \frac{x_{max} - x_{min}}{\sqrt{n}}$$

If $n < 25$ the width of the class interval can be determined better according to Sturges' rule:

$$\Delta x = \frac{x_{max} - x_{min}}{1 + 3.3 \log n}$$

If too small an interval width is chosen too small, a jagged 'envelope' of the bar graph will result. If too large an interval width is chosen, the 'envelope' is quantised too coarsely and the shape of the graph remains hidden.

A bar graph can be normalised by plotting $N(x)/n$ instead of $N(x)$. The vertical axis then represents the *fraction* of measurements that lie within a given interval. We can restate this as follows: 'The *probability* of a measurement result lying within a given interval is now plotted along the vertical axis'. In addition, one can also normalise the bar graph with respect to the interval width Δx, by plotting $N(x)/n\Delta x$ instead of $N(x)/n$. After normalisation a bar graph is usually referred to as a *histogram*.

If the number of samples n increases and the range $x_{max} - x_{min}$ remains a bound interval, as is the case in practice for all physical quantities, then the number of intervals or bars of the histogram will increase, whereas the width Δx will decrease. As $n \to \infty$, the envelope of the histogram will tend towards a smooth curve. This (double) normalised histogram is called the *probability density function* $f(x)$. It is defined by:

$$f(x) = \lim_{\Delta x \to 0} \frac{1}{n} \frac{N(x)}{\Delta x}$$

We can also write:

$$f(x) \, dx = \frac{N(x)}{n}$$

This implies that $f(x) \, dx$ is the probability of a sample scoring between x and $x + dx$: hence the name *probability density function*. From the last expression it follows that:

$$\int_{-\infty}^{\infty} f(x) \, dx = 1$$

This integral represents the sum of all probabilities f(x) dx. This is equal to the probability of the sample scoring either in the first interval dx, or in the second, or in the third, and so on. Since the measured sample x must lie in one of these intervals, the sum must equal 1. The last expression shows that the area under a probability density function is equal to unity (which is basically achieved by normalising twice). From a probability density function $f(x)$ such as the one in Fig. 2.14 one can easily read the probability of a measured sample x being less than a certain value a. This probability, which is denoted by $P(x < a)$ can be expressed as:

$$P(x < a) = \int_{-\infty}^{a} f(x)\, dx$$

This is precisely the area under $f(x)$ to the left of the line $x = a$ (see Fig. 2.14).

In the same way, for a given density function $f(x)$, the mean $\bar{x}$ of a collection of sample values x_i can be written as:

$$\bar{x} = \lim_{n \to \infty} \frac{1}{n} \sum_{i=1}^{n} x_i = \int_{-\infty}^{\infty} x\, f(x)\, dx$$

The variance can be expressed as:

$$\sigma_x^2 = \lim_{n \to \infty} \frac{1}{n-1} \sum_{i=1}^{n} (x_i - \bar{x})^2 = \int_{-\infty}^{\infty} (x - \bar{x})^2\, f(x)\, dx$$

And again, the standard deviation is the square root of the variance:

$$\sigma_x = \sqrt{\int_{-\infty}^{\infty} (x - \bar{x})^2\, f(x)\, dx}$$

Figure 2.14. Probability density function.

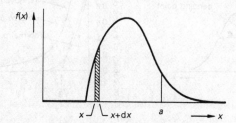

If the errors in the measured samples are caused by a large number of mutually independent events, they can be proven to be distributed according to a particular probability density function: the *normal* or *Gaussian* density function. The proof is comprised in the central limit theorem of the theory of probability. Random measurement errors are generally assumed to have a normal distribution, although for many cases this is not true.

The normal probability density function is given by:

$$f(x) = \frac{1}{\sigma\sqrt{2\pi}} \exp\left\{-\frac{(x - \overline{x})^2}{2\sigma^2}\right\}$$

and has the shape plotted in Fig. 2.15(a). The probability of $x < \overline{x} - a$ or $x > \overline{x} + a$ leads to the following type of integrals:

$$\int_{-\infty}^{\overline{x}-a} f(x)\,dx$$

and

$$\int_{\overline{x}+a}^{\infty} f(x)\,dx$$

which cannot be determined analytically.

A numerical approximation of the solution is given in Table 2.1.

When a measurement is afflicted with random errors, one cannot define a finite interval around the measurement result for which we can be absolutely sure that it contains the true value of the measured quantity. The

Figure 2.15. (a) Normal or Gaussian probability density function. (b) Probability of a measurement deviating more than $k\sigma$ from the mean $\overline{x}$.

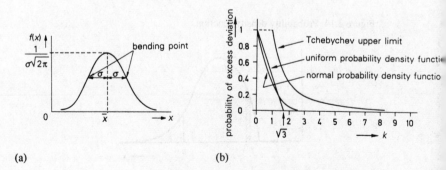

(a)   (b)

expression 'maximum possible error' therefore only applies to systematic errors. Even so, an error interval is often given for measurements exhibiting random errors. There is, however, a non-zero probability that the true value lies outside this interval: the 'probability of excess deviation'. Obviously, this probability depends on the width of the error interval with respect to the width of the probability density function. This probability increases as the error interval becomes narrower and/or the density function becomes wider. However, we can find an upper limit for this probability. This upper limit is independent of the shape of the probability density function and holds for all measurements afflicted with random errors and a given (finite) mean $\bar{x}$ and standard deviation σ. This upper limit is given by the inequality of Tchebychev–Bienaymé:

$$P\{|x - \bar{x}| \geq k\sigma\} \leq \frac{1}{k^2} \qquad (k > 0 \text{ and real})$$

This inequality asserts that there is an upper limit $1/k^2$ to the probability of a measurement x, afflicted with random errors, deviating more than a certain value $k\sigma$ from the mean $\bar{x}$. For $0 < k \leq 1$ the expression becomes trivial.

To prove this inequality we recall the expression for the variance:

$$\sigma^2 = \int_{-\infty}^{\infty} (x - \bar{x})^2 f(x) \, dx$$

$$\sigma^2 \geq \int_{-\infty}^{\bar{x}-k\sigma} (x - \bar{x})^2 f(x) \, dx + \int_{\bar{x}+k\sigma}^{\infty} (x - \bar{x})^2 f(x) \, dx \quad (k \geq 0)$$

The inequality is immediately evident with:

$$\int_{\bar{x}-k\sigma}^{\bar{x}+k\sigma} (x - \bar{x})^2 f(x) \, dx \geq 0 \quad \text{if} \quad k \geq 0$$

Table 2.1. *The probability that a Gaussian distributed measurement value x with mean $\bar{x}$ and standard deviation σ lies outside 1σ, 2σ and 3σ intervals around the mean value.*

x outside the interval	Probability
$(\bar{x} - \sigma, \bar{x} + \sigma)$	0.32
$(\bar{x} - 2\sigma, \bar{x} + 2\sigma)$	0.045
$(\bar{x} - 3\sigma, \bar{x} + 3\sigma)$	0.0026

The term $(x -\bar{x})^2 \geq k^2\sigma^2$ holds when $x \leq \bar{x} - k\sigma$, but also when $x \geq \bar{x} + k\sigma$, $(x - \bar{x})^2 \geq k^2\sigma^2$. Consequently, we may replace the term $(x -\bar{x})^2$ in the integrals by the smaller term $k^2\sigma^2$, and obtain:

$$\sigma^2 \geq k^2\sigma^2 \left\{ \int_{-\infty}^{\bar{x}-k\sigma} f(x)\, dx + \int_{\bar{x}+k\sigma}^{\infty} f(x)\, dx \right\}$$

This last expression is exactly equal to:

$$k^2\sigma^2\, P\{|x - \bar{x}| \geq k\sigma\} \quad (k \geq 0),$$

thus, if $k > 0$, one can conclude:

$$P\{|x - \bar{x}| \geq k\sigma\} \leq \frac{1}{k^2} \quad \text{QED}$$

In Fig. 2.15(b) the upper limit of the probability of x exceeding the interval $(\bar{x} - k\sigma, \bar{x} + k\sigma)$ is plotted as a function of k. If we *know the density function* $f(x)$, we can calculate the probability of x exceeding the interval with the following method, rather than using the above estimate for the upper limit, because Tchebychev–Bienaymé always yields a pessimistic approximation of the upper limit.

Expanding the left-hand term of the inequality gives:

$$P\{|x - \bar{x}| \geq k\sigma\} = \int_{-\infty}^{\bar{x}-k\sigma} f(x)\, dx + \int_{\bar{x}+k\sigma}^{\infty} f(x)\, dx$$

This can be rewritten as:

$$1 - \int_{\bar{x}-k\sigma}^{\bar{x}+k\sigma} f(x)\, dx, \quad \text{since} \quad \int_{-\infty}^{\infty} f(x)\, dx = 1$$

The *uniform* probability density function is described by:

$$f(x) = \frac{1}{b - a} \quad \text{for } a \leq x \leq b$$

$$f(x) = 0 \quad \text{elsewhere}$$

It is found that for this probability density function $\bar{x} = (a + b)/2$ and $\sigma = (b - a)/2\sqrt{3}$; the probability of excess deviation equals $1 - k/\sqrt{3}$, which is also drawn in Fig. 2.15(b). The probability of excess deviation for the case of a normal distribution is also plotted in Fig. 2.15(b) (with the aid of a look-up table). Therefore, once we know the probability density function, we can calculate the probability of excess deviation. However, even when this function is unknown, we can still provide an upper limit! For a normal distribution, for instance, the probability of x exceeding the mean by $\pm$ 2.56σ ($k = 2.56$) is 1%. If we do not know the probability density function, we can only state a much larger error interval, i.e. $\pm 10\sigma$ ($k = 10$). This is caused by lack of information about the shape of the probability density function for that specific measurement process.

If the expression 'maximum possible error' is used, one normally refers to a deviation of no more than $\pm 3\sigma$, which means in only 0.28% of the cases is the error actually larger than this value. A probability of only 0.28% is small enough to be disregarded for most cases, although, strictly speaking, we cannot speak of a 'maximum limit' for a random error. Furthermore, if we consistently use $\pm 3\sigma$ errors in our calculations, the uncertainty of an end result, which relies on a number of individual measurements (each of which is afflicted with random errors) will become excessively large. In the next section on error propagation, we will explain that the variance σ^2 is a more useful parameter for characterising random errors.

Sometimes the expression 'probable random error' is used. By this, we mean the width of the interval that contains exactly half of all collected samples. For random errors with a normal distribution this error is equal to $\pm 0.67\sigma$.

Fig. 2.16 shows the result of a measurement the samples of which are afflicted with both random and systematic errors. The random errors are distributed according to a normal probability density function here. The true value of the measured quantity is a. A systematic error causes the mean of the samples to be shifted to b. The total error (with a probability of excess deviation of 0.14%) is equal to the sum of the systematic error ($a - b$) and

Figure 2.16. Random and systematic errors.

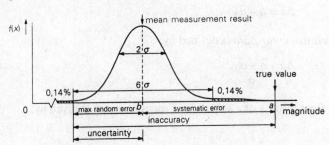

the 'maximal random error'. This total error determines the *inaccuracy* of the measurement. The *uncertainty* of a measurement is a measure of the discrepancy between the samples, due only to random errors. The uncertainty usually corresponds to the 3σ interval of the probability density function of the random errors. Strictly speaking the uncertainty of a measurement is given by an interval within which the true value of the measurand lies with a given confidence probability.

The *repeatability* of a measurement is a measure for the extent of agreement between consecutive measurements of the same physical quantity, using the same method and equipment, and under the same operating conditions over a short period of time. The repeatability is determined by the uncertainty in the measurement. A small uncertainty gives a large repeatability. The *reproducibility* of a measurement is the closeness of agreement among repeated measurements of the same measurand performed in different locations under different operating conditions, or spread out over a long period of time. As this can introduce different systematic errors, the reproducibility is generally worse than the repeatability.

2.3.2 Error propagation

Often, the final result x of a sequence of measurements is determined by performing mathematical operations on a number of measured individual physical quantities a, b, c, ..., so $x = f(a,b,c,...)$. The errors in a, b, c, ..., will, of course, contribute to the error in x. The analysis of the sensitivity of x to the errors in a, b, c, ... is called *error sensitivity analysis* or *error propagation analysis*. We will discuss this analysis first for systematic errors, then for random errors.

Systematic errors

Let a be the result of an individual measurement and a_0 the true value of the measured physical quantity, then the *absolute error* Δa is defined by:

$$\Delta a = a - a_0$$

The *relative error* $\Delta a/a$ is defined by:

$$\frac{\Delta a}{a} = \frac{a - a_0}{a}$$

If the final result x of a sequence of measurements is given by:

$$x = f(a,b,c,...)$$

in which a, b, c, ... are the values of physical quantities with absolute errors Δa, Δb, Δc, ..., respectively, then the absolute error Δx of the final result x is given by:

$$\Delta x = f(a,b,c,...) - f(a - \Delta a, b - \Delta b, c - \Delta c,...)$$

With a Taylor expansion of the second term, this can also be written as:

$$\Delta x = \frac{\partial f(a,b,c,...)}{\partial a} \Delta a + \frac{\partial f(a,b,c,...)}{\partial b} \Delta b + ...$$

in which all higher-order terms have been neglected. This is permitted provided that the absolute errors Δa, Δb, Δc, ... are small, and also that the higher order partial derivatives of $f(a,b,c,...)$ are small. In other words, if the curvature of $f(a,b,c,...)$ at the point $(a,b,c,...)$ is small, and the environment of interest around the point $(a,b,c,...)$ is small (determined by Δa, Δb, Δc, ...) only the slope of $f(a,b,c,...)$ is important.

One never knows the actual value of Δa, Δb, Δc, ... Usually the individual measurements are given as $a \pm \Delta a_{max}$, $b \pm \Delta b_{max}$, ..., in which Δa_{max}, Δb_{max} are the maximum possible errors. In this case, for the maximum possible error Δx_{max} in the end result the following holds:

$$\Delta x_{max} = \left| \frac{\partial f(a,b,c...)}{\partial a} \right| \Delta a_{max} + \left| \frac{\partial f(a,b,c,...)}{\partial b} \right| \Delta b_{max} + ...$$

If the individual contributions in the right half of this expression are all equal, we call this measurement optimal. NB The measurement is optimal with respect to the relative contribution of all individual error sources. An optimal measurement does not necessarily give the lowest possible overall error. The expression above can be rewritten to obtain the maximal *relative* error:

$$\left| \frac{\Delta x_{max}}{x} \right| = \left| \frac{\partial f}{\partial a} \frac{a}{f} \frac{\Delta a_{max}}{a} \right| + \left| \frac{\partial f}{\partial b} \frac{b}{f} \frac{\Delta b_{max}}{b} \right| + ...$$

Using the following notation:

$$\frac{\partial f}{\partial a} \frac{a}{f} = S_a^x; \quad \frac{\partial f}{\partial b} \frac{b}{f} = S_b^x \quad \text{etc.}$$

this becomes:

$$\left| \frac{\Delta x_{max}}{x} \right| = |S_a^x| \left| \frac{\Delta a_{max}}{a} \right| + |S_b^x| \left| \frac{\Delta b_{max}}{b} \right| + \dots$$

The factors S_a^x, S_b^x, ... are called the *sensitivity factors* of the end measurement x to the relative errors in the measured values $a, b, \dots$.

From the definition of the sensitivity factor we can deduce several rules of computation which simplify the error sensitivity analysis considerably. Some of these rules are:

1. $S_{a_1 \cdot x_2}^x = S_{a_1}^x + S_{a_2}^x$ 2. $S_a^{x^n} = n S_a^x$ 3. $S_{a^n}^x = n^{-1} S_a^x$

4. $S_{a^{-n}}^{x^n} = -S_a^x$ 5. $S_{a^n}^{x^n} = S_a^x$ 6. $S_a^x = S_b^x S_a^b$

The sensitivity factors provide a means for easy determination of how systematic errors will propagate through sums, products, ratios, etc. of measurement results. If the errors in the measured values $a, b, c, \dots$ were all maximal and their signs were such that they amplify one another, the above expressions would yield the true error in x, rather than the maximum possible error.

Random errors
As we have seen, in the case of random errors we cannot really speak of the 'maximum possible error'. If we were to continue to measure for a long enough period of time, an error will occur, which is larger than the largest error in all previous measurements. Therefore, for the analysis of the propagation of random errors in a series of measurements, it is better to use the mean value $\bar{a}$ and the variance σ_a^2 of the individual measurements. Let us assume that the end result x again depends on a number of individually measured physical quantities $a, b, c, \dots$, so:

$$x = f(a,b,c,\dots)$$

Further we also know the values of $\bar{a}, \bar{b}, \bar{c}, \dots$ and $\sigma_a^2, \sigma_b^2, \sigma_c^2, \dots$, which have been calculated from a number of samples of the quantities $a, b, c, \dots$. We can than write:

$$\bar{x} = f(\bar{a},\bar{b},\bar{c},\dots)$$

and

$$\sigma_x^2 = \left(\frac{\partial f}{\partial a} \right)^2_{(\bar{a},\bar{b},\bar{c},\dots)} \sigma_a^2 + \left(\frac{\partial f}{\partial b} \right)^2_{(\bar{a},\bar{b},\bar{c},\dots)} \sigma_b^2 + \left(\frac{\partial f}{\partial c} \right)^2_{(\bar{a},\bar{b},\bar{c},\dots)} \sigma_c^2 + \dots$$

Together, these two expressions form *Gauss' error propagation rule*. With this rule we can determine the mean $\bar{x}$ and the variance σ_x^2 of the end result

x, which is a function of a number of individual measurements a, b, c, ... afflicted with random errors.

The proof for this rule goes as follows. Small deviations da, db, dc, ... in the means $\bar{a}$, $\bar{b}$, $\bar{c}$, ... cause a deviation in x dx, which amounts to:

$$dx = \left(\frac{\partial f}{\partial a}\right)_{(\bar{a},\bar{b},\bar{c},...)} da + \left(\frac{\partial f}{\partial b}\right)_{(\bar{a},\bar{b},\bar{c},...)} db + \left(\frac{\partial f}{\partial c}\right)_{(\bar{a},\bar{b},\bar{c},...)} dc + ...$$

Again, we have neglected the higher order terms of the Taylor expansion, as we did in the case of systematic errors. Consequently, this expression is only valid for small deviations and a small curvature of the function $x = f(a, b, c, ...)$ at the point $\bar{a}$, $\bar{b}$, $\bar{c}$, ... Since dx is the deviation of x_e from $\bar{x}$, the variance σ_x^2 is equal to the average value of $(dx)^2$, which we will denote as $\overline{(dx)^2}$. Then, σ_x^2 can be expressed as:

$$\sigma_x^2 = \overline{(dx)^2} = \overline{\left(\frac{\partial f}{\partial a} da + \frac{\partial f}{\partial b} db + ...\right)^2}$$

Expanding, this results in:

$$\sigma_x^2 = \overline{\left(\frac{\partial f}{\partial a}\right)^2 (da)^2 + \left(\frac{\partial f}{\partial b}\right)^2 (db)^2 + ... + \frac{\partial f}{\partial a}\frac{\partial f}{\partial b} da\, db + ...}$$

$$\underbrace{\qquad\qquad\qquad\qquad\qquad}_{\text{squares}} \qquad \underbrace{\qquad\qquad\qquad}_{\text{cross products}}$$

For random errors, the individual errors da, db, dc, ... will generally be uncorrelated, causing the cross products to be as likely to be positive as negative and to have positive values similar to the negative values. This results in an averaged value of zero for these cross products. Thus:

$$\overline{\frac{\partial f}{\partial a}\frac{\partial f}{\partial b} da\, db + ...} = 0$$

If we denote $\overline{(da)^2} = \sigma_a^2$; $\overline{(db)^2} = \sigma_b^2$, ... the expression for σ_x^2 can be written as:

$$\sigma_x^2 = \left(\frac{\partial f}{\partial a}\right)^2_{(\bar{a},\bar{b},\bar{c},...)} \sigma_a^2 + \left(\frac{\partial f}{\partial b}\right)^2_{(\bar{a},\bar{b},\bar{c},...)} \sigma_b^2 + \left(\frac{\partial f}{\partial c}\right)^2_{(\bar{a},\bar{b},\bar{c},...)} \sigma_c^2 + ...$$

QED

In the above derivation of Gauss' error propagation rule, the shape of the probability density function of the individual measurements a, b, c, ... did not matter. This rule, therefore, also applies to random errors which are not normally distributed.

For instance, we can apply Gauss' rule to the case in which the end result x of a measurement is equal to the average of n individual measurements a_i ($i = 1, ..., n$). Therefore:

$$x = \frac{1}{n} \sum_{i=1}^{n} a_i$$

If we denote the mean of all possible observations a_i as $\overline{a}$, then:

$$\overline{a} = \lim_{n \to \infty} \frac{1}{n} \sum_{i=1}^{n} a_i$$

and the variance σ_a^2 of all observations a_i becomes

$$\sigma_a^2 = \lim_{n \to \infty} \frac{1}{n-1} \sum_{i=1}^{n} (a_i - \overline{a})^2$$

Applying Gauss' rule results in:

$$\overline{x} = \overline{a}$$

for the mean value of the end result x, and

$$\sigma_x^2 = \frac{1}{n^2} n \sigma_a^2 = \frac{1}{n} \sigma_a^2$$

or

$$\sigma_x = \frac{\sigma_a}{\sqrt{n}}$$

for the variance of the end result x. The last expression shows that the measurement certainty, when taking the average of n individual measurements (with random errors) of the same physical quantity, increases with the square root of the number of measurements.

2.3.3 Sources of error

In order to reduce measurement errors as much as possible, it is desirable to acquire some insight into their sources. We will now therefore investigate these error sources based on the measurement chain depicted in Fig. 2.17.

In a measurement chain consisting of the measurement object, the measurement system and the observer in a particular environment, the object and the system interact; the object will influence the measurement system, and *vice versa*, the system will affect the object. The first form of interaction is a desired one. It realises the transfer of measurement information. The latter form of interaction is undesirable. The object can be affected in such a way and to such an extent that the measured quantity changes noticeably. This constitutes an *influencing error*. One must try to match the input stage (hatched in Fig. 2.17) of the measurement system to the measurement object in such a way that this influencing error is reduced to a minimum. Section 2.3.3.1 deals with errors that result from measurement systems influencing the measurement object. It also deals with the *matching* necessary for reducing this influence.

Fig. 2.17 also illustrates the interaction between the output of the measurement system and the observer. The expression 'observer' is used here in a wide sense of the word which covers more than just a human observer. The 'observer' may be human, but can also be a machine. For instance, when the measurement result is used to control a machine or process that machine/process constitutes the 'observer'. Again, the interaction of measurement system and observer consists of a desirable and an undesirable component. The influence of the measurement system on the observer, i.e. the transfer of the result to the observer, is obviously the desired component. The reverse influence can result in an influencing error, if the observer acts upon the measurement system in such a way and to such an extent as to alter the measurement result. Examples of this kind of error are: reading a dial at an angle, rather than straight, resulting in a parallax error; loading of a measurement system output caused by a mismatched (artificial) observer. In order to reduce *influencing errors* we must adapt the output stage of the measurement system to the observer. This is discussed in Section 2.3.3.4.

Besides the two interactions mentioned above, the measurement system

Figure 2.17. Interactions of a measurement system with the object under test, the environment and the observer.

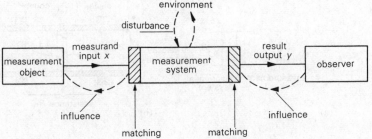

also interacts with its environment. If this occurs in such a way and to such an extent that the environment actually affects the measurement result, the interaction is undesirable and is called perturbation or interference. This source of measurement errors is referred to as *disturbance* or 'interference' and will be dealt with in Section 2.3.3.3.

Finally, a fourth source of measurement errors arises from the (imperfect) characteristics of the measurement system itself. If the system characteristics are inadequate for the measurement required, they will cause incorrect measurements. Several important characteristics which can be used to minimise these measurement errors (*system errrors*) are discussed in Section 2.3.3.2. The characteristics include bandwidth, response time, etc.

2.3.3.1 Influencing the measurement object: matching

The mere act of measuring always affects the measurement object to a certain degree. It may cause the measured quantity to change more than we bargain for. To avoid this one must match the measurement system to the measurement object. Usually it is sufficient to alter only the input portion (stage) of the measurement system. According to the situation, one can distinguish various kinds of matching. We will only discuss three of these in detail.

Anenergetic matching

Anenergetic matching is designed to reduce to a minimum the transfer of energy or power between the measurement object and the measurement system. After matching, measurement will not supply (any appreciable) energy to, or receive from the measurement object.

Fig. 2.18 shows the principle of anenergetic matching for measuring an across quantity V_b. The read-out of the measurement system is proportional to V_i, which is given by:

Figure 2.18. Example of anergetic matching of an across quantity: the measurment of flow viscosity using a turbine emter. (a) Network analogy of the flow measurement. (b) Turbine flow meter.

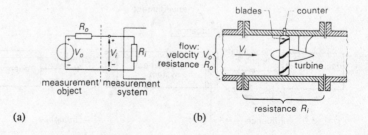

(a)   (b)

$$V_i = V_o \frac{R_i}{R_i + R_o}$$

When the measurement system is connected, the (generalised) voltage at the terminals of the object V_o will drop to V_i. This is an undesired influence by the measurement system. Assuming α is the permissible relative inaccuracy in the measurement of V_o, R_i of the measurement system must satisfy:

$$R_i \geq \frac{R_o}{\alpha} - R_o \approx \frac{R_o}{\alpha}$$

For an accurate measurement of an across quantity, we conclude that the internal resistance of the measurement system must be much larger than that of the measurement object. In that case the power supplied by the object will be very small. Furthermore, this power will be dissipated almost entirely in R_i of the measurement system and not in R_o of the object. This does not seem to hold for the air flow measurement of Fig. 2.18(b), as R_i must become infinitely large. However, with the analogy method (see Section 5.4), we find that $R_i = 1/D_i$, in which D_i is the damping introduced by the turbine. Hence, the damping of the turbine must be as small as possible.

When a measured object may not be burdened, we must attempt anenergetic matching. For measurement of an across quantity this implies that the input impedance of the measurement system must be much larger than the source impedance of the object. If the complex input impedance of the measurement system has a (non-zero) positive real value, the measurement system will always absorb (some) energy. If the input impedance is purely imaginary, on average, the system will not absorb energy, but there will be an instantaneous exchange of energy. Examples of this kind of measurement situation are a capacitive voltmeter and the U-shaped manometer.

The following is an example of anenergetic matching for measuring a through quantity (see Fig. 2.19). The deflection or reading of the measurement system is proportional to I_i:

$$I_i = I_o \frac{R_o}{R_o + R_i}$$

Because $R_i \neq 0$, $I_i \neq I_o$. If the permissible relative inaccuracy is again equal to α, then:

$$R_i \leq \alpha R_o$$

Therefore, for an accurate measurement of a through quantity, the internal resistance of the measurement system must be far smaller than that of the measured object. The object delivers almost no power and what little power dissipation occurs takes place virtually entirely within the measurement system (in R_i).

Fig. 2.19(b) depicts the measurement of a mechanical force. For the sake of simplicity, the measurement object has been omitted here: F_i is the force at the input of the measurement system. This situation is analogous to Fig. 2.19(a), if the R_o and R_i are replaced by two inductors, L_o and L_i. The measurement object generates a force F_o (internally), which must compress a spring with a force spring $K_o = 1/L_o$. The stiffness of the Bourdon gauge and that of the suspension of the pressure plate give rise to a spring constant of the measurement system of $K_i = 1/L_i$. Clearly, the voltage between the terminals in Fig. 2.19(a) and the velocity of the pressure plate in Fig. 2.19(b) are small when $L_i \ll L_o$ or when $K_i \gg K_o$. In this case the system will measure $F_i \approx F_o$.

Thus we can generalise that for the measurement of a through quantity, good anenergetic matching is achieved when the input impedance of the measurement system is much smaller than the internal impedance of the measured object. If the input impedance is purely imaginary, as for instance in Fig. 2.19(b), the average delivered power is zero, but the instantaneous power is not zero. Therefore, the observations made above for across quantities also apply to these measurements. Other examples of the measurement of through quantities and are an electro-dynamic ammeter and a force gauge.

The internal impedance Z_o of a measurement object can be calculated with the aid of Thévenin's theorem (see Fig. 2.20). Let V_o be the open circuit voltage of the object and I_s the short circuit current, then:

Figure 2.19. Example of anergetic matching of a through quantity: the measurment of a mechanical force by means of a Bourdon gauge. (a) Network analogy of the force measurement. (b) Pressure plate force meter.

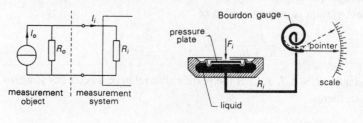

(a)   (b)

$$Z_o = \frac{V_o}{I_s}$$

Norton's theorem can often be useful for determining the equivalent circuit of a measurment object (see Fig. 2.20).

Energic matching

The aim of this type of matching is to extract the maximum available power from the measured object, so that the required power gain in the measurement system can be smaller.

Energic matching is especially important for passive measurement systems, e.g. measurement systems which do not possess internal power amplification. It is easy to determine the conditions which the input impedance Z_i of such a system must satisfy in order to maximise the average power P_{avg} delivered into the measurement system, for a given measurement object. Since, on average, the imaginary part of the input impedance $Z_i = R_i + jX_i$ does not absorb any power, the following expression holds (see Fig. 2.21(a)):

$$P_{avg} = I^2 R_i$$

If the impedance of the measured object is given as $Z_o = R_o + jX_o$, the RMS-value of the current is equal to:

$$I = \frac{V_o}{\sqrt{(R_o + R_i)^2 + (X_o + X_i)^2}}$$

Figure 2.20. Thévenin's and Norton's theorems.

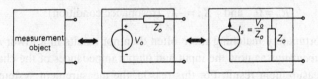

Figure 2.21. Energic matching of (a) a measurement system to an object, and (b) an object to a measurement system.

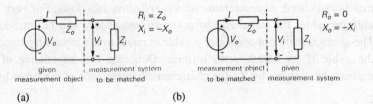

(a)   (b)

and, therefore, the power P_{avg} which is delivered to Z_i is given by:

$$P_{avg} = \frac{V_o^2 R_i}{(R_o + R_i)^2 + (X_o + X_i)^2}$$

This power is maximal for a given object with V_o and Z_o, if:

$$R_i = R_o \quad \text{and} \quad X_i = -X_o$$

This last condition is the so-called 'resonance condition'.

A given measurement object, with a (generalised) internal impedance Z_o, will deliver most power if the input impedance of the measurement system is chosen to be:

$$Z_i = Z_s^*$$

where Z_s^* is the conjugate impedance of Z_s. Therefore, the maximum available power a (measurement) object can deliver to a (measurement) system is:

$$P_{avg,max} = \frac{V_o^2}{4R_i} = \frac{V_o^2}{4R_o}$$

In this expression V_o is the RMS-value of the open circuit (generalised) voltage of the measurement object. The power transferred in the case of perfect energetic matching is called the *available power*.

Note that Fig. 2.19(b) shows that it is also possible (in some cases) to adapt the measurement object to an existing measurement system. It follows that, for this case, maximum power is delivered when:

$$R_o = 0 \quad \text{and} \quad X_o = -X_i \text{ (resonance condition)}$$

Unfortunately matching more often than not results in a lower *measurement accuracy*, as now the input and output impedances of the chain affect the measurement result. For this reason, the measurement systems almost always used are active systems (with built-in power gain).

For transporting high-frequency measurement signals along transmission lines from a measurement object to a measurement system, another form of matching is used: *characteristic* or *non-reflective matching*. For very long uniform cables the so-called *characteristic impedance* can be introduced. The characteristic impedance of a cable is equal to the input impedance of the cable if its length were infinite. Denoting the impedance of the measurement object by Z_o, the characteristic impedance of the cable by Z_c, and the input impedance of the measurement system by Z_i, characteristic matching is achieved when:

$$Z_s = Z_o = Z_i$$

If this condition is satisfied, no reflections will occur at the ends of the cable (see Fig. 2.22).

For a lossless cable, the characteristic impedance is given by:

$$Z_c = \sqrt{\frac{L}{C}} = R_c$$

where L is the series inductance per metre of the cable conductors and C is the parallel capacitance per metre between the conductors. The characteristic impedance of a lossless cable has only a real part and no imaginary part. Reflectionless matching requires therefore $R_o = R_c = R_i$. When this holds, however, energic matching is also achieved simultaneously, since $R_o = R_i$. The characteristic resistance R_c of the cable is only an *apparent resistance* which does not dissipate any energy (as long as the cable is lossless). Therefore, half of the energy delivered by V_o is dissipated in R_o, and the other half in R_i of the measurement system.

If a transmission line is not terminated characteristically, reflections off the ends of the line will cause standing waves on the line; the output signal is no longer a good measure for the input signal.

Table 2.2. *The characteristic impedance of several connections.*

Type of connection	Characteristic impedance
coaxial cable	50–75 Ω
printed circuit board traces	50–150 Ω
twisted wire pairs	100–120 Ω
ribbon cable (2.5 cm)	200–300 Ω
free space	376 Ω

Figure 2.22. Reflectionless matching occurs when $Z_o = Z_c = Z_i$.

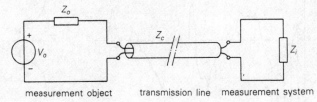

measurement object   transmission line measurement system

Noise matching

In every physical system which dissipates energy (converting the energy to heat) small randomly fluctuating signals will occur. Since it is impossible, strictly speaking, to make lossless physical systems, the following applies to all physically realisable systems. The randomly fluctuating signals originating in these systems are called noise, in analogy to the hissing sound that is produced when these signals are made audible. Every practical physical system has an energy efficiency less than 100% — it will always dissipate some energy — and therefore, at a temperature higher than absolute zero, every physical system is afflicted with noise. This omnipresent thermal noise is induced by the thermal agitation of the charge carriers in conductors and by particles such as atoms and molecules in Brownian motion.

Every measurement will add its own noise to the measured signal. Since this signal is extracted from a physical system, it will itself already contain noise. The purpose of *noise matching* is to let the measurement system add as little noise as possible to the measurand. We will first see how we can characterise a measurement signal afflicted with thermal noise.

Fig. 2.23 shows a constant voltage of time averaged magnitude $\overline{x}$, with noise superimposed on it. The time averaged value of the noise is zero. The probability density function of the noisy signal has been drawn on the right-hand side. In order to determine the RMS-value of this signal we take n samples x_i ($i = 1, ..., n$). The RMS-value is defined as the Root of the Mean of the Squares (RMS) of the samples. Let us consider the noise $(x_i - \overline{x})$ in the measured n samples around the average value $\overline{x}$. The RMS-value of the noise only is given by:

$$x_{RMS} = \sqrt{\frac{1}{n}\sum_{i=1}^{n}(x_i - \overline{x})^2}$$

Figure 2.23. Measurement signal with noise $x(t)$ and amplitude probability density function $f(x)$.

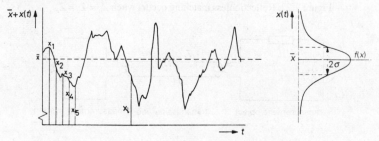

NB We have subtracted $\bar{x}$ from the sample values x_i: this expression resembles the expression for the standard deviation σ of n samples x_i round a mean $\bar{x}$ which we found earlier. If n is sufficiently large, the RMS-value x_{RMS} of the noise will asymptotically approach the standard deviation σ of the samples:

$$x_{\mathrm{RMS}} = \sigma_x$$

If we take the limit for the number of samples going to infinity ($n \rightarrow \infty$) and the time between samples approaching zero ($\Delta t \rightarrow 0$), the above expression will also apply to continuous non-sampled noise: the RMS-value of the noise is therefore equal to the standard deviation of the noise. From this, it immediately follows that the average power P_{avg} of a noisy signal is proportional to its variance σ^2; $P_{\mathrm{avg}} = V^2/R = I^2 R$, in which V and I are RMS-values.

When two noisy voltages sources are connected in series, or two noisy currents are added, the resulting voltage or current, respectively is given by:

$$x(t) = a(t) + b(t) \quad (\textit{superposition of instantaneous values})$$

With Gauss' rule for error propagation and $x = f(a,b)$, $\partial f/\partial a = 1$ and $\partial f/\partial b = 1$, we find:

$$\sigma_x^2 = \sigma_a^2 + \sigma_b^2 \quad (\textit{superposition of power})$$

This means we can add the power or the instantaneous value of two noisy signals $a(t)$ and $b(t)$ to represent the sum of both signals, provided that the fluctuations in these signals are *uncorrelated*. The last statement is a necessary condition only for the validity of Gauss' rule. Thus it applies only to adding the power of two noise signals to arrive at the power of the sum signal. If we express the power summation in RMS-values we get:

$$x_{\mathrm{RMS}} = \sqrt{a_{\mathrm{RMS}}^2 + b_{\mathrm{RMS}}^2}$$

Normally, noise covers a wide frequency band. The power of the noise within a given narrow band Δf is usually not the same for every frequency (see Fig. 2.24). The noise is called 'white' if the distribution across the frequency spectrum is uniform (see mid-section of Fig. 2.24). The noise is called 'pink' if the noise per unit of bandwidth decreases with increasing frequency (see the left-hand section of Fig. 2.24).

An important form of noise is *thermal noise*. As we have seen this noise is caused by thermal agitation of, for instance, charge carriers in a resistor, but also by the Brownian motion of particles in a (generalised) mechanical resistance. A capacitance microphone will, for instance, produce noise due to the Brownian motion of the molecules surrounding its membrane, although the capacitive electrodes (capacitor) are noiseless (or their resistance would be zero). It appears that all (dissipating) generalised resistors produce noise (including the non-electrical ones). We can model a noisy resistor by a noiseless resistor in series with a noise voltage source, or as a noiseless resistor in parallel with a noise current source (see Fig. 2.25). The RMS-values of the noise of the voltage and the current source are respectively:

Figure 2.24. An example of the spectral distribution of noise and interference in a practical measurement system.

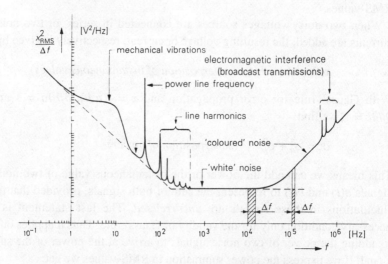

Figure 2.25. Resistor noise.

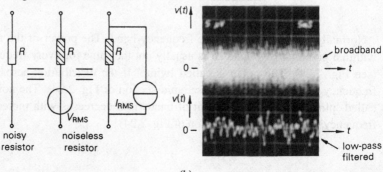

(a)   (b)

$$V_{RMS} = \sqrt{4kTR\Delta f} \quad \text{and} \quad I_{RMS} = \sqrt{\frac{4kT\Delta f}{R}}$$

In these expressions k is Boltzmann's constant ($k = 1.38 \times 10^{-23}$ J/K) and T is the absolute temperature. The only part of a complex impedance Z producing noise is the real part. Therefore, more general expressions are:

$$V_{RMS} = \sqrt{4kT\Delta f \, \text{Re}(Z)}$$

and

$$I_{RMS} = \sqrt{4kT\Delta f \, \text{Re}(Y)} = \sqrt{4kT\Delta f \, \text{Re}\left(\frac{1}{Z}\right)}$$

Ideal capacitors and inductors therefore do not produce noise since they are lossless. The power of the noise signal of a resistor (proportional to V_{RMS}^2 or I_{RMS}^2) in a frequency band Δf is constant; the thermal noise of a resistor is *white noise*.

Deriving the expression for the thermal noise V_{RMS} of a resistor is relatively straightforward if we follow a train of thought first suggested by Nyquist. The 'gedanken experiment' is illustrated in Fig. 2.26. A resistor R, in thermal equilibrium with its surroundings, is connected to each end of a lossless transmission line of length l. The characteristic impedance Z_c of this line is matched to R, so no reflection will occur in the line. The equivalent noise $V_{1,RMS}$ of the left resistor produces a voltage $\frac{1}{2} V_{1,RMS}$, which will propagate along the line to the right at a speed v_c in the form of a 'voltage wave'; the corresponding 'current wave' equals $\frac{1}{2} V_{1,RMS}/R$. In the same manner, the right resistor produces a voltage wave and a current wave which propagate to the left. If we now close both switches S at the same time, we will have a piece of lossless cable which is short-circuited at each end. This effectively constitutes a resonator, with a resonance frequency: $f_n = nv_c/2l$. Assuming the switches and the cable are ideal, we can assert that in both situations (S open and S closed) the average energies of the waves in

Figure 2.26. Derivation of the thermal noise of a resistor. $V_{1,RMS} = V_{2,RMS}$.

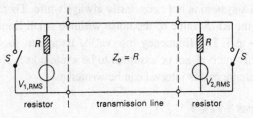

the cable are equal. The number of cable resonance frequencies that lie within a frequency band Δf is equal to $\Delta f/(v_c/2l) = 2l\Delta f/v_c$.

Each resonance frequency is associated with two degrees of freedom: one for the electrical and one for the magnetic energy (or, in the case of a mechanical system, kinetic and potential energies). According to the law of equipartition of energy the average energy, per degree of freedom is $\frac{1}{2}kT$. Therefore, the total average energy in the cable in a frequency band Δf is equal to $(2l/v_c)kT\Delta f$. When the switches are open, each resistor R will deliver an average energy equal to $(2l/v_c)V_{RMS}^2/4R$ in $2l/v_c$ seconds into the cable. Obviously, this energy is dissipated by the other resistor since the cable is lossless and the resistors are in thermal equilibrium with their environment. Furthermore, on average, this energy is equal to the energy that is stored in the travelling wave in the cable when the switches are closed. Combing the last two expressions results in:

$$\frac{2l}{v_c}\frac{V_{RMS}^2}{4R} = \frac{2l}{v_c}kT\Delta f$$

from which follows:

$$V_{RMS} = \sqrt{4kTR\Delta f} \quad \text{QED}$$

Thermal noise is not the only source of noise in a measurement system. The fact that mass and charge are quantised causes noise when there is a flow of charge or mass across a (generalised) potential barrier. This noise is called *shot noise*. The reason for this noise is that the particles which pass the potential barrier have a certain kinetic energy probability density function. Only the fastest particles will actually cross the barrier, so only the right-hand tail of the probability density function contributes to the particle flow. Thus, the flow of particles, the number of particles per unit of time, is not constant, but exhibits random fluctuations. Obviously the relative effect of this I_{RMS}/I is largest when the flow or current is small. This leads to the expression $I_{RMS} = \sqrt{2qI\Delta f}$ for shot noise, where q is the charge or mass of the mass carriers (particles) that result in the current (flow) I.

The noise in a system is not necessarily always white. Therefore, we will now consider the RMS-value of the noise within a small limited frequency interval $(f, f + \Delta f)$. The frequency interval Δf is chosen so small that the noise within this interval can be assumed to be white noise. The RMS-value of the noise voltage in this interval can be written as:

$$V_{RMS} = V_n\sqrt{\Delta f}$$

in which V_n is the RMS-value of the noise in a 1 Hz wide frequency band. The square root in the expression can be accounted for by remembering that the power of white noise is proportional to $V_{RMS}^2 = V_n^2 \Delta f$. The voltage V_n is called the *spectral noise voltage density*. In the same manner the RMS-value I_{RMS} of the noise current within a frequency band Δf is given by:

$$I_{RMS} = I_n \sqrt{\Delta f}$$

Here I_n is the spectral noise current density for a bandwidth of 1 Hz. V_n and I_n are often referred to as spot noise or slot noise.

Note that since the units of V_{RMS} and I_{RMS} are volt and ampere, respectively, the physical dimensions of V_n and I_n must be $[V_n] = V/\sqrt{Hz}$ and $[I_n] = A/\sqrt{Hz}$.

A measure for the extent to which a signal can be distinguished from the background noise is the *signal-to-noise ratio*:

$$S/N = \frac{\text{signal power}}{\text{noise power}}$$

In this expression the signal power is defined as the power which is dissipated in the load of the signal source (in the input of the measurement system) and the noise power is defined as the power dissipated in the same (noiseless) load. Both powers are considered within the same small frequency interval $(f, f + \Delta f)$. Usually, the signal-to-noise ratio is frequency dependent.

Assuming that the noise of a measurement object is equal to the thermal noise of (the real part of) its internal impedance Z_o, and the object is loaded with a noiseless impedance Z_i, the power dissipated in Z_i can be calculated as follows. The noise voltage is given by:

$$V_{RMS} = \sqrt{4kT\Delta f \, \text{Re}(Z_o)}$$

The noise power dissipated in (the real part of) Z_i is:

$$\frac{V_{RMS}^2 \, \text{Re}(Z_i)}{|Z_o + Z_i|^2}$$

The signal power dissipated in Z_i is:

$$\frac{V_o^2 \, \text{Re}(Z_i)}{|Z_o + Z_i|^2}$$

where V_o is the RMS-value of the source voltage of the measurement object. Thus, the signal-to-noise ratio becomes:

$$S/N = \frac{V_o^2}{4kT} \, \Delta f \, \text{Re}(Z_o)$$

The signal-to-noise ratio is sometimes defined as the ratio of the *available signal power* and the *available noise power*. This is the power which is dissipated in the load impedance if it is chosen equal to the conjugate of the internal impedance of the signal source. However, this arbitrary choice does not change the magnitude of the signal-to-noise ratio. Here it is only important that both signal power and noise power are referred to the same impedance.

As we mentioned before, a measurement system adds noise to the signal. A measure for the worsening of the noise is the *noise figure*:

$$F = \frac{N_o}{N_o'}$$

Here, N_o is the noise power at the *output* of the noisy measurement system, with the noisy measurement signal connected to the output. N_o' is the noise power at the output of the same system, which is now considered to be free of noise. The output noise then comes only from the measured input signal.

Let us define the power amplification G of a measurement system as:

$$G = \frac{S_o}{S_i}$$

in which S_o is the signal power — within a frequency interval $(f, f + \Delta f)$ — which dissipates in the system output load impedance, and S_i is the signal power which dissipates in the system input impedance. This yields:

$$F = \frac{N_o}{S_o} \frac{S_o}{N_o'} = \frac{N_o}{S_o} \frac{S_i G}{N_i G} = \frac{S_i/N_i}{S_o/N_o}$$

where N_i is the noise power of the measured object that dissipates in the input impedance of the measurement system. Thus, the noise figure F also shows the deterioration of the signal-to-noise ratio caused by the measurement system.

We can also use the equivalent input noise power $N_{oi} = N_o/G$ to calculate the total noise N_o at the output of the measurement system. With $N_i = N_o'/G$ we find:

$$F = \frac{N_o}{N_o'} = \frac{N_{oi}}{N_i}$$

N_{oi} is the noise power which is required at the input of a noiseless measurement system to produce the noise power N_o that is observed at the output.

The noise of a measurement system can be represented by two spot noise sources at the input. To this end, we assume the measurement system to be free of noise and we place an equivalent noise voltage source and an equivalent noise current source at the input terminals of the system, such that they together produce the same noise at the output of the system (see Fig. 2.27). We may assume that the two input noise sources are uncorrelated. This is a practical assumption for all but extremely high frequencies. In addition, we assume the noise to be white across the entire bandwidth B of the system. Then the voltage source will produce $V_n\sqrt{B}$ and the current source: $I_n\sqrt{B}$, where V_n and I_n are spectral noise densities. V_n is determined by measuring the RMS-value of the output noise $AV_n\sqrt{B}$ with the input short-circuited. The equivalent current noise source then cannot contribute to the system output noise. If we know the gain A and the bandwidth B, we can calculate V_n. If the RMS-value $AI_nR_i\sqrt{B}$ of the output noise is also determined with the input terminals open, with known values of A, B and R_i, we can compute I_n. Again with the system input terminals open, the equivalent input voltage noise source cannot contribute to the system output noise.

As far as the noise of the measurement object is concerned, we assume in the following that it is white and has a spectral voltage density V_R. It should be mentioned that this noise is not necessarily equal to the thermal noise produced by the internal impedance R_o of the measurement object, but it can be larger, due to the presence of other, non-thermal, noise sources.

We have already seen that the noise figure F can be written as:

$$F = \frac{N_o}{N_o'} = \frac{S_i/N_i}{S_o/N_o} = \frac{N_{oi}}{N_i}$$

Figure 2.27. A noisy measurement system with its equivalent circuit.

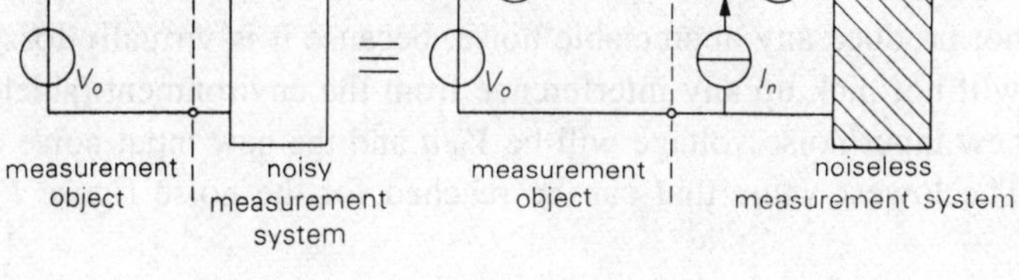

measurement object | noisy measurement system $\equiv$ measurement object | noiseless measurement system

The noise power of the measurement object dissipating in the input resistance R_i of the measurement system is given by:

$$N_i = \left(\frac{V_R R_i}{R_o + R_i}\right)^2 \frac{1}{R_i} = \frac{V_R^2 R_i}{(R_o + R_i)^2}$$

The noise power of the equivalent input noise sources V_n and I_n dissipating in R_i amounts to:

$$\frac{V_n^2 R_i}{(R_o + R_i)^2} + \frac{I_n^2 R_o^2 R_i}{(R_o + R_i)^2}$$

Therefore, the total equivalent noise power N_{oi} at the input of the measurement system is:

$$N_{oi} = N_i + \frac{R_i}{(R_o + R_i)^2} \{V_n^2 + I_n^2 R_o^2\}$$

The noise figure F thus becomes:

$$F = \frac{N_{oi}}{N_i} = 1 + \frac{V_n^2 + I_n^2 R_o^2}{V_o^2}$$

Apparently, the minimal value for F is 1, when the system does not add any noise at all and V_n and I_n become zero. Unfortunately, such a system is not realistic. We must therefore noise match the measurement system to the measurement object (which has a given V_R and R_o) so that the noise figure is as close to unity as possible. This means that the term $V_n^2 + I_n^2 R_o^2$ must be as small as possible. One way of doing this is to give the first stage of the measurement system a large gain so that the noise of the following stages becomes negligible. Other often used means involve choosing the correct low-noise amplifying devices for the input and the optimal biasing of these devices. If no further improvement can be achieved by changing the measurement system input, the following two methods can be used to lower F further.

This can be accomplished this by connecting a transformer (or, with non-electrical system inputs, the equivalent of a transformer) between the measurement object and the system. The input of the transformer (with a turns ratio of 1 : n) becomes the new input of the system. The transformer will not produce any appreciable noise, because it is virtually lossless (it also will not pick up any interference from the environment (shielding)). The new input noise voltage will be V_n/n and the new input noise current nI_n. The lowest value that can be reached for the noise figure F when varying n is now attained when $V_n/I_n = n^2R_o$. Clearly when $V_n/I_n = R_o$ we do not need the transformer: the measurement object is already noise matched to the system input.

The second method involves connecting j input stages, each with their own V_n and I_n, in parallel and taking the sum of the individual outputs divided by j as the output. This combination of input stages will have an equivalent input noise of $V_n/\sqrt{j}$ and $I_n\sqrt{j}$. Noise matching is accomplished here when $V_n/I_n = jR_s$. For *both* methods the noise figure becomes:

$$F = 1 + 2\frac{V_nI_nR_o}{V_R^2}$$

We can conclude from this result that when one uses one or both of these methods one must minimise the product V_nI_n rather than $V_n^2 + I_n^2R_o^2$ to reach the smallest possible value of F.

The previous discussion is based on a given measurement object in which the only manipulable is the (input of the) measurement system. However, if the measurement object may also be changed we must first maximise the signal-to-noise ratio $S_i/N_i = V_o^2/V_R^2$ before minimising F by changing the measurement system. In conclusion we have learned that *noise matching* of the measurement system to a given measurement object is reached if:

$$v_n/I_n = R_o$$

In that case no further improvement of the noise figure F is possible by the means discussed above.

2.3.3.2 Characteristics of measurement mystems
In this section we will examine several characteristics of measurement systems which can influence the correctness of the measurement result. If one or more of these characteristics does not correspond to a required value (the specified value) errors will occur in the measurement results.

Sensitivity
The *sensitivity* S of a (linear) measurement system is the ratio of the magnitude of the output signal y to that of the input signal x

$$S = \frac{y}{x}$$

The sensitivity of a measurement system is generally frequency dependent: $S = \overline{S}(\omega)$.

The sensitivity of a measurement amplifier is usually called the *gain*, whereas for a (measurement) system we generally speak of the *transfer function*. Apart from the sensitivity, the *scale factor W* is sometimes used. This is defined as:

$$W = \frac{1}{S} = \frac{x}{y}$$

Compare: The height of the grid on the screen of an oscilloscope is 8 cm. The electron beam is deflected across the entire grid by an oscilloscope input signal with a peak-to-peak value of 40 mV. Then the sensitivity S is 0.2 cm/mV and the scale factor W is 5 mV/cm. It is this scale factor that is usually given for oscilloscopes.

When the transfer relationship $y = f(x)$ between the output signal y (the reading) and the input x (the quantity to be measured) is non-linear, we cannot speak of a sensitivity since the ratio of the output y and the input x varies with the magnitude of x. For these non-linear systems we introduce the *differential sensitivity*. The differential sensitivity S_{diff} of a measurement system, characterised by $y = f(x)$, is defined as:

$$S_{\mathrm{diff}}(x_0) = \left(\frac{\mathrm{d} f(x)}{\mathrm{d}x} \right)_{x=x_0}$$

for an input signal x_0. For a linear system $S_{\mathrm{diff}} \neq S_{\mathrm{diff}}(x_0)$ and $S_{\mathrm{diff}} = S$. For a non-linear system S_{diff} depends on value of the input x.

Take, for example, a null detector with a non-linear transfer function $y = ax - bx^3$, in which $a > 0$ and $b > 0$. This has a decreasing differential sensitivity for an increasing input signal. It is especially important for a null detector to have a high differential sensitivity for very small input signals. The larger $S_{\mathrm{diff}}(0)$, the better the zero situation can be detected and the more accurate the measurement can be.

Another measure for the sensitivity of a non-linear measurement system is the *sensitivity factor*. The sensitivity factor S_x^y of a measurement system output y for an input x is defined as (the transfer relationship is $y = f(x)$):

$$S_x^y = \frac{\dfrac{\mathrm{d}y}{y}}{\dfrac{\mathrm{d}x}{x}} = \frac{\mathrm{d} f(x)}{\mathrm{d}x} \cdot \frac{x}{y}$$

The notation S_x^y is interpreted as the sensitivity factor of y to variations in x. In the case of a linear system, S_x^y is not a good measure of sensitivity, since, irrespective of the magnitude of S, $S_x^y = 1$.

We have already come across the sensitivity factor in our discussion of measurement error propagation (Section 2.3.2). Another example of the use the sensitivity factor in metrology is a strain gauge. This transducer converts a variation of the length Δl to a change in resistance ΔR. The sensitivity factor S_l^R of the strain gauge is given by:

$$S_l^R = \frac{\dfrac{\Delta R}{R}}{\dfrac{\Delta l}{l}} = \frac{dR}{dl} \cdot \frac{l}{R}$$

Note that the system sensitivity S discussed above is dimensionless only when y and x have the same physical dimension. This is never the case, for instance, for transducers. The differential sensitivity S_{diff} has the same dimension as the sensitivity S. The sensitivity factor, however, is always dimensionless.

Sensitivity threshold

There is no point in increasing the sensitivity of a measurement system indefinitely (for instance, by increasing the gain): eventually the *sensitivity threshold* of the system will be reached.

The sensitivity threshold of a measurement system is determined by the smallest input signal which can still be detected, with a given probability of success. The sensitivity threshold prevents us from detecting arbitrarily small signals. This is due to the fact that in every realisable physical system spontaneous, random fluctuations (noise) will occur, causing a small input measurement signal to 'drown' in the (background) noise. The noise in a measurement system can have many causes, such as thermal agitation (resistor noise) or the quantised character of the flow of charge, mass or energy carriers (shot noise of electrons, ions or photons) across a potential barrier.

Besides unavoidable fundamental sources of noise there are other disturbances in a measurement system which can obscure the true signal. For example, mechanical vibration or electrical interference may cause such a large output signal that small input signals can no longer be distinguished. Mechanical defects, such as friction, play or a dead zone, can give rise to a threshold for the sensitivity, below which an input signal will not produce an output signal. Often, only by constructing the measurement system differently can these non-fundamental limits be eliminated.

The fundamental limit to the system sensitivity as a meaningful characteristic is determined by the random fluctuations within that system. The theoretically practicable sensitivity threshold is determined by the noise which is always present in a measurement system.

We will analyse the sensitivity threshold of a noisy measurement system in the case of a constant measurand value x. Assume the noise has a Gaussian probability density function. Therefore, when $x = 0$, the noise of the output signal y will have a density function $f_n(y)$ with $\bar{y} = 0$ (see Fig. 2.28). If an input signal $x \neq 0$ is connected, the output will consist of both the desired signal $\bar{y}$, and (the same) noise. The density function is now denoted $f_{sn}(y)$.

The important question is now: what amplitude of $\bar{x}$ can just be detected? In other words, which mean $\bar{y}$ can still be distinguished from the situation when $x = 0$, and therefore $\bar{y} = 0$? The answer to this question depends on the certainty with which we wish to know whether or not there is a signal hiding in the noise. This can easily be understood following the reasoning below. Let us assume the *RMS-value* σ of the noise of the output (the standard deviation of $f_n(y)$) to be equal to $1/n$th of the output signal $\bar{y}$. Due to polarity symmetry of the probability density function $f_n(y)$ we can introduce a *detection criterion* based on whether or not the actual output y is smaller or larger than $\frac{1}{2}\bar{y}$. Imagine y is a sampled output. We must then be able to make a statement on the existence of a signal $\bar{y}$, based on only one sample of y. (If, however, we are allowed to postpone our conclusion and take the average of several samples, we in fact apply low-pass filtering. The probability of detection then increases considerably, since the effective signal-to-noise ratio increases.) If $y > \frac{1}{2}\bar{y}$ we conclude that the input signal exists, but if $y < \frac{1}{2}\bar{y}$ our conclusion is that there is no input. Fig. 2.28(b) shows the situation in which a great number of samples has been taken, both with and without an input signal. Here n is about 3 ($\bar{y} \approx 3\sigma$). The dark band between both noise afflicted output traces disappears when $n = 2$. Then we can no longer clearly distinguish the two traces and we get the two density functions as depicted in Fig. 2.28(a). What is the *detection certainty* in this case? For $\bar{y} = 2\sigma$ ($v = 2$), we can see from Fig. 2.28(a) that (for a detection criterion larger or smaller than $\frac{1}{2}\bar{y}$) the conclusion of 'no input signal' will be incorrect for 16% of the samples. This is precisely that section of the total area under $f_{sn}(y)$ which is cross hatched. Therefore, the chance of detecting an input signal x which causes an output signal $\bar{y} = 2$ is 84%. So the *detection* probability is 84%. Thus, it is possible to detect a DC voltage with a certainty of 84%, when this voltage is obscured by noise with an RMS-value which is equal to half the value of the DC voltage ($n = 2$). The signal-to-noise ratio is $(n\sigma)^2/\sigma^2 = n^2 = 4$. This demonstrates that the desired detection certainty determines the sensitivity threshold (and therefore n).

The detection certainty or detection probability of the existence of an input signal, for the criterion $y > \frac{1}{2}\bar{y}$, has been calculated for several values of $\bar{y}$ and tabulated in Table 2.3.

A commonly used measure for the sensitivity threshold is the magnitude of the input signal for which the signal-to-noise ratio is equal to unity. The detection probability is then approximately 70% for noise with a normal amplitude probability density function.

The previous discussion assumes that we wish to base our conclusion about the presence of an input signal on only one simple sample or

Table 2.3. *The detection probability and the signal-to-noise ratio for different values of the signal y, in relation to the standard deviation σ of the noise in the signal.*

Signal $\bar{y}$	Detection probability	*S/N* ratio
1σ	69.15%	1
1.4σ	76.11%	2
2σ	84.13%	4
3σ	93.32%	9
4σ	97.72%	16
5σ	99.38%	25
6σ	99.87%	36
8σ	99.9968%	64
10σ	99.999971%	100

Figure 2.28. The sensitivity threshold of a measurement system afflicted with noise. (a) The sytem output probability density function with no input signal applied, $f_n(y)$, and the output density function when an input signal x is applied giving an output signal, $f_{sn}(y)$. (b) The output signal as a function of time for the no input signal case ($x = 0$) and the case in which an input signal is applied resulting in an output signal $\bar{y}$.

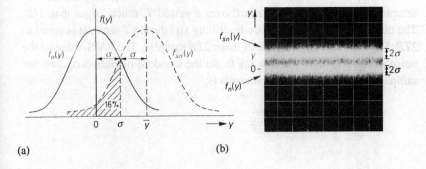

(a)   (b)

measurement. The sensitivity threshold improves when we base our conclusion on a number of samples (say n). We have already seen that:

$$\sigma_{\text{avg}} = \frac{\sigma}{\sqrt{n}}$$

in which σ is the RMS-value of the noise and σ_{avg} is the standard deviation of the average of n samples. Thus, by averaging, the signal-to-noise ratio is increased by a factor n and the sensitivity threshold improves correspondingly.

The sensitivity threshold can also be improved by reducing the bandwidth B of the measurement system. Assuming the noise to be white, the RMS-value σ of the noise is given by:

$$\sigma = \sigma_0 \sqrt{B}$$

where σ_0 is the equivalent noise in a frequency band of 1 Hz. This means that if the bandwidth B of a measurement system is reduced by a certain factor (and the input signal remains the same), the signal-to-noise ratio will improve by the same factor. The sensitivity threshold again improves accordingly.

Besides taking the average of n individual successive samples, we can also measure the input signal $x(t)$ continuously for a particular time T. The time average y_{avg} of the measurement systems' output $y(t)$, for the interval $(t, t + T)$ is:

$$y_{\text{avg}} = \frac{1}{T} \int_{t}^{t+T} y(t)\, dt$$

We can now use this averaged value to establish whether or not there is a signal at the input. To determine the resulting improvement of the sensitivity threshold, we use Shannon's sampling theorem: If in a signal $y(t)$ there are no frequencies higher than B Hz, the signal is completely defined by samples taken $1/2B$ seconds apart, over a period T, much larger than $1/B$. The number of discrete samples describing $y(t)$ during T seconds is equal to $2TB$. We now take the average of these $2TB$ samples. The RMS-value of the noise in the signal $y(t)$ is $\sigma = \sigma_0 \sqrt{B}$. So the standard deviation σ_{avg} of the sample average of $y(t)$ over T seconds is:

$$\sigma_{\text{avg}} = \frac{\sigma}{\sqrt{2TB}} = \frac{\sigma_0}{\sqrt{2T}}$$

Taking the average over a time T therefore increases the signal-to-noise ratio by a factor of $2T$; the sensitivity threshold improves by a factor of $1/\sqrt{2T}$.

Summarising, we can state that the sensitivity threshold is the smallest signal one can still detect with a particular certainty, above the background noise of the measurement system. The sensitivity threshold depends on the required detection certainty and the amplitude of the noise of the measurement system. We can reduce the noise by using a measurement system with a smaller bandwidth, or by determining the average value over a number of individual measurement samples, or by taking the time average of a continuous measurement over a time T. All these measures take a longer time before the result is known; they make the response of the measurement system slower. This is the penalty we pay for lowering the intrinsic sensitivity threshold of a measurement system.

Signal shape sensitivity

The input signal of a measurement system is the carrier of the information on the value of the physical quantity which is to be measured. The response of the system to the input signal is generally dependent on the shape (form or structure) of the input signal.

Often, signals are classified as follows: A signal may be constant, i.e. independent of time (*static signals*), for instance, a constant voltage or a DC current. Usually the signal will vary only slowly and is therefore called *quasi-static*. Often, though, the signal will be time dependent (*dynamic signals*). If the signal $x(t)$ repeats in time every T seconds, it is called a *periodic signal*, with period T (for all t, $x(t) = x(t + T)$, see Fig. 2.29(a)). The repetition frequency is $f = 1/T$. For pulse shaped signals as in Fig. 2.29(b), the ratio $\Delta t/T$ is defined as the *duty cycle*. Signals with a very small duty cycle (impulses) are difficult to measure and are often the cause of strong interference in neighbouring measurement set-ups. When a time dependent signal has no periodicity, it is called a *unique* or *transient signal*. Examples of this are noise voltages and transient phenomena, such as overshoot and

Figure 2.29. Two examples of a periodic measurement signal. (a) ECG-signal: the period here is $T = 1$ s. (b) An impulse signal: the duty cycle here is $\Delta T/T = 1/7$.

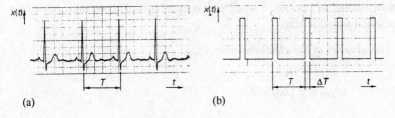

(a)   (b)

damped ringing of switching systems. In general, periodic signals are easier to measure than non-periodic signals. Therefore, when measuring, for example, the transient response of a system, the response is repeated many, many times. In this way, a periodic signal which is easy to measure is obtained.

A dynamic signal can be analysed in the *time domain* and also in the *frequency domain*. Measurement of the *structure* of the wave form with, for instance, an oscilloscope is performed in the time domain, whereas measurement of the frequency content is done with a wave or spectrum analyser in the frequency domain.

For a dynamic signal we must therefore specify to which characteristic of a signal the measurement system responds in order to define the concept of 'sensitivity of the measurement system' unambiguously. In other words: Which *signal characteristic value* is being measured? For a dynamic measurement signal $x(t)$ one can define the following characteristic signal values:

– The *peak value* x_p:

$$x_p = \max |x(t)|$$

– The *peak-to-peak value* x_{pp}:

$$x_{pp} = \max \{x(t)\} - \min \{x(t)\}$$

It is recommended that the peak and peak-to-peak values be used as little as possible, since they are both very sensitive to disturbances such as noise which are superimposed on the true signal. Distortion in the signal will also easily result in large errors in x_p and x_{pp}. The following signal values are far less sensitive to distortion and interference:

– The *time average value* x_{avg}:

$$x_{\text{avg}} = \frac{1}{T} \int_{t}^{t+T} x(t)\, dt$$

The average of a periodic signal is computed for an interval which is a whole number of periods, $T = n/f$ with n an integer. The average value of a sinusoidal signal is zero.

– The *average of the absolute value*, $|x|_{\text{avg}}$:

$$|x|_{\text{avg}} = \frac{1}{T} \int_{t}^{t+T} |x(t)|\, dt$$

When referring to the average of a sinusoidal signal, one usually means the average of the absolute value of this sine wave.

– The *RMS-value* x_{RMS}:

$$x_{RMS} = \left[\frac{1}{T} \int_t^{t+T} x(t)^2 \, dt \right]^{\frac{1}{2}}$$

The following shows the usefulness of characterising a measurement signal by its RMS-value: The instantaneous power $p(t)$ which is dissipated in a resistor R with the measurement signal $x(t)$ across it is:

$$p(t) = \frac{v(t)^2}{R} = v(t)i(t) = \frac{v(t)^2}{R} = i(t)^2 R$$

Here $i(t)$ is the resulting current through the resistor R. The average power P_{avg} which dissipates during a time interval T in the resistor is given by:

$$P_{avg} = \frac{1}{T} \int_t^{t+T} p(t) \, dt$$

$$= \frac{1}{TR} \int_t^{t+T} v(t)^2 \, dt = \frac{V_{RMS}^2}{R}$$

$$= \frac{R}{T} \int_t^{t+T} i(t)^2 \, dt = I_{RMS}^2 R$$

Hence, as long as we use the RMS-values of measurement signal voltage and current, we can easily calculate the (average) dissipated power. Obviously, it is also possible to define the RMS-value of a voltage or current as being equal to the value of a DC voltage or DC current that would generate the same amount of heat in a resistor as the voltage or current to be measured (heat definition).

The design of a measurement system determines to which characteristic value of a signal the system responds. A measurement system may react to the instantaneous value of a signal (oscilloscope), to the average value (moving coil meter), to the average of the absolute value (moving coil meter with a rectifier/amplifier) or to the RMS-value (electro-dynamic voltmeter).

For the sinusoidal signal $x(t) = a \sin(\omega t)$ in Fig. 2.30, the characteristic values can be calculated to be:

$$x_p = a; \qquad x_{pp} = 2a \qquad |x|_{avg} = \frac{2a}{\pi} \,; \qquad x_{RMS} = \frac{a}{\sqrt{2}}$$

The ratio $x_{RMS}/|x|_{avg}$ is called the *form factor* of the signal $x(t)$. The ratio x_p/x_{RMS} is called the *crest factor* of the signal. For a sine wave the form factor equals 1.11 and the crest factor $\sqrt{2}$. The form factor is important when the measurement system measures the average of the absolute value, although it is calibrated to indicate the RMS-value (for a sine wave). This is the case for many electronic voltmeters. The crest factor is important in measurement of noise and impulse shaped signals. Often we wish to know the RMS-value of such signals, but we must make sure the peak value of the signal stays within the linear range of the system to avoid measurements errors due to saturation.

Resolution
The resolution of a measurement system is a measure of the detail for which the system can be adjusted, or of the detail which can be read-out of the system. The resolution is defined as the smallest interval Δx of the value x of the measured quantity that will still cause a change in the measurement result y. The resolution R is given numerically by:

$$R = \frac{x}{\Delta x}$$

Sometimes the maximum value of the resolution is stated. This occurs for the maximum value of x that can be measured with a system without saturation, distortion or overload:

$$R_{max} = \frac{x_{max}}{\Delta x}$$

The resolution R has a finite value for all systems in which the measurement result does not increase continuously with the measured

Figure 2.30. The peak-to-peak value x_{pp}, the RMS-value x_{RMS} and the average of the absolute value $|x|_{avg}$ of a sine wave signal.

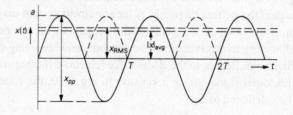

quantity x. For instance, in a mechanical measurement system with play or stiction, in a wire-wound position potentiometer, in a step-attenuator and in numerical displays the output value y does not continuously increase with the input, but in small steps Δy.

When the system resolution is finite, the measurement becomes quantised; we will make a *quantisation error*. Errors that are the result of measurement quantisation can be divided into *truncation errors* and *round-off errors*. A truncation error is made when the system simply omits the decimal places to the right of the lowest indicated decimal place: the remainder. This usually occurs in α-numerical displays, for instance a digital voltmeter. The magnitude of the error is here $-\Delta x/x$; the smallest possible step Δx divided by the displayed value x. A round-off error is made when the lowest indicated decimal place takes into account the remainder by rounding off to the nearest decimal place. Here the error is equal to $+\Delta x/2x$; i.e. half the smallest step divided by the displayed value. If, for a null measurement, we use a reference which can be adjusted only in steps, and we set the reference to the value for which the null detector shows the smallest reading, we will be making a round-off error.

Non-linearity

The relation between the output signal y and the input signal x of a measurement system with a frequency independent sensitivity is linear when $y = f(x)$ is a straight line, so when $f(x) = ax + b$ such systems with a frequency independent sensitivity are called *static systems*. The relation between $y(t)$ and $x(t)$ for a linear, frequency dependent system is given by a linear differential equation. This is an equation which contains only first order terms of y and derivatives of y. Systems with a frequency dependent sensitivity are called *dynamic systems*. If the sum of two sinusoidal signals $x_1 = a_1 \sin(\omega_1 t + \phi_1)$ and $x_2 = a_2 \sin(\omega_2 t + \phi_2)$ is connected to the input of a linear measurement system, the output will also consist of two sinusoidal signals $y_1 = b_1 \sin(\omega_1 t + \phi_{o1})$ and $y_2 = b_2 \sin(\omega_2 t + \phi_{o2})$, with the same frequencies as the input signal (*isochronism*), but different amplitude and phase.

For linear systems the *principle of superposition* holds, this means that if an input signal x_1 produces an output signal y_1 and an input x_2 results in an output y_2, the linear combination $ax_1 + bx_2$ as input will produce an output of $ay_1 + by_2$. The superposition principle only applies to the component of the system output signal that is caused by the input; possible offset errors are not taken into consideration here.

If the sum of two sine waves of different frequencies is applied to a non-linear system, the output signal will contain *harmonics*. Harmonics are sine waves with frequencies $n\omega_1$ and $k\omega_2$, respectively, in which n and k are integers. $n, k = 1$ gives the fundamental harmonic, higher values for n and k produce the higher nth and kth order harmonics. In addition, the output will also contain signal components which have the *sum and difference frequencies $n\omega_1 \pm k\omega_2$*, where $\omega_1 \geq \omega_2$ and $n \geq k$. Besides this, the output may also contain a *DC component*.

This can easily be seen using the Taylor expansion of the relation $y = f(x)$ of a static, non-linear system at, for instance, the point $x = 0$:

$$y = f(0) + x f'(0) + \frac{x^2}{2!} f''(0) + \dots$$

With:

$$f(0) = 0, \quad f'(0) = c_1, \quad f''(0) = c_2 \ \text{ etc.}$$

and:

$$x(t) = x_1(t) + x_2(t)$$

in which:

$$x_1(t) = a_1 \sin \omega_1 t \text{ and } x_2(t) = a_2 \sin \omega_2 t$$

we find:

$$y(t) = \frac{c_2}{4}(a_1^2 + a_2^2) + c_1(a_1 \sin \omega_1 t + a_2 \sin \omega_2 t)$$
$$- \frac{c_2}{4}(a_1^2 \cos 2\omega_1 t + a_2^2 \cos 2\omega_2 t) - \frac{c_2}{2}\{a_1 a_2 \cos(\omega_1 + \omega_2)t$$
$$- a_1 a_2 \cos(\omega_1 - \omega_2)t\} + \dots$$

The degree of non-linearity of a measurement system is characterised by the *non-linear* or *harmonic distortion*. This distortion is measured with a single sine wave input to the measurement system. The magnitude of the distortion is defined as the ratio between the RMS-value y_n of the nth harmonic and the RMS-value y_1 of the fundamental (or first) harmonic. The distortion due to the nth harmonic is defined as:

$$d_n = \frac{y_n}{y_1} \quad (n > 1)$$

The total distortion of $n - 1$ harmonics follows as (excluding the fundamental ($n = 1$)):

$$d = \frac{y_t}{y_1} = \sqrt{d_2^2 + d_3^2 + \ldots d_n^2}$$

This relation will be clear if we realise that the RMS-value of all $n - 1$ higher harmonics is the root of the sum of the squares of the RMS-values:

$$y_t = \sqrt{y_2^2 + y_3^2 + \ldots + y_n^2}$$

The degree of *static* (frequency independent) *non-linearity* is often specified in a different manner. Assume that the true relation between y and x is given by $y = f(x)$ as illustrated in Fig. 2.31. The best linear approximation of the curve $y = f(x)$ is given by $y = ax$. A measure of the non-linearity of the measurement system can now be given as the maximal value of the expression:

$$\left| \frac{f(x) - ax}{ax} \right|$$

over the entire dynamic range of the measurement system. This definition of non-linearity is also referred to as the zero-biased non-linearity.

No measurement system will be perfectly linear; it will only be approximately linear (for instance, over a small range of the input x). We will now briefly discuss the several forms of non-linearity that one can encounter in practical measurement systems. First of all we will show some forms of static non-linearity.

– *Saturation and clipping*. This non-linearity is characterised by a *decreasing* differential sensitivity S_{diff} with an increasing input x. As shown in Fig. 2.32 the decrease occurs abruptly in the case of clipping, and gradually with saturation. To prevent damage to a measurement

Figure 2.31. Static non-linearity.

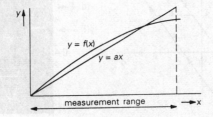

system when the system is overloaded by too large an input signal, the output swing is often deliberately limited to safe values.

– *Hysteresis* This form of non-linearity causes a different curve in the *xy*-plane for increasing values of *x* than it does for decreasing values (see Fig. 2.33). Hysteresis can be brought about, for instance, by play in a mechanical gear drive. Another example of hysteresis is the relation between the magnetic induction *B* and the field strength *H* in a ferromagnetic material (*B-H* curve).

– *Dead band.* Typical for this form of non-linearity is the existence of one (or more) regions in which the output signal amplitude *y* is independent of the input signal *x* (see Fig. 2.34). This non-linearity can be caused by static friction (stiction). An object (such as a pointer) will not move until the exerted force is larger than the static friction.

Besides the forms of static non-linearity, discussed above, any possible combinations of these can also occur. In addition, there are dynamic (frequency dependent) forms of non-linearity. One of these, for instance, is

Figure 2.32. Example of saturation and clipping in measurement systems. (a) Non-linear input *x* to output transfer cuves. (b) Output signals *y*(*t*) as a result of a sinusoidal input signal *x*(*t*).

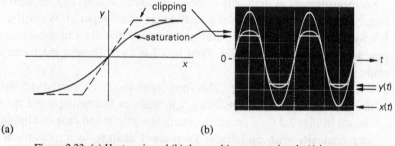

(a)   (b)

Figure 2.33. (a) Hysteresis and (b) the resulting output signal *y*(*t*) in response to a sinusoidal input signal *x*(*t*).

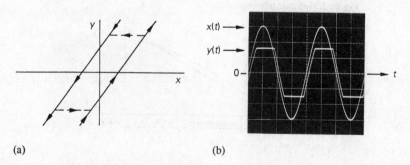

(a)   (b)

slew-rate distortion, which we will illustrate with the example of an oscilloscope. If an oscilloscope amplifier can supply only a maximum current I_{max} to the deflection electrodes and the deflection electrodes have a capacitance C, then the rise and fall time of a displayed wave form cannot be smaller than a certain minimum value. This minimum is determined by:

$$\frac{dV_c}{dt} = \frac{I_{max}}{C}$$

in which V_c is the voltage across the electrodes, proportional to the deflection of the electron beam. Thus the time derivative of V_c is proportional to the tracing speed of the beam along the screen. Fig. 2.35 shows that as soon as either the frequency or the amplitude of the input signal becomes so large that the maximum for dV_c/dt is exceeded, the displayed wave will be distorted to a triangle.

Figure 2.34. (a) Dead band in the input–output transfer curve of a system. (b) The output response $y(t)$ to a sinusoidal input $x(t)$.

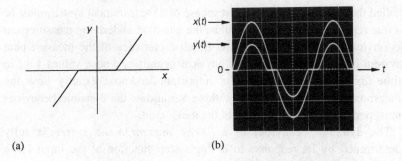

(a)   (b)

Figure 2.35. Slew rate distortion: $x(t)$ is the input isgnal and $y(t)$ the output signal.

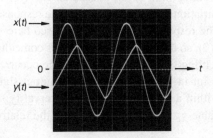

Measurement range, dynamic range

The measurement range of a system is determined by the interval (x_{min}, x_{max}) within which the system can still measure the input quantity with the specified accuracy. The *dynamic range* of the measurement system is equal to the ratio x_{max}/x_{min}. Usually x_{max} is determined by the maximum limit of the allowable non-linearity, which is reached for large input signals. The magnitude of x_{min} is usually determined by zero-bias errors and noise which, relatively speaking, become increasingly significant for smaller signals. x_{min} is the smallest value of x for which a given accuracy can be achieved.

Take, for instance, the case in which a current must be measured with an inaccuracy of ±3%, with a moving coil meter with an inaccuracy of ±1% full scale. The dynamic range is in this case only 3. Another example is when a voltage must be measured to a ±1% inaccuracy with an instrumentation amplifier. The zero offset of the amplifier referred to the output is less than 10 μV. Due to non-linearity, the input voltage must remain below 10 V, to achieve an inaccuracy of 1%. The dynamic range is 10^3 in this example.

System response

The reaction of a measurement system to a stimulus applied to the input is called the *system response*. The response of a measurement system must be a true representation of the stimulus; the intention underlying measurement is obviously not also to measure the characteristics of the measurement system itself! When measuring physical quantities whose values vary in time (*dynamic quantities*), it is important to know exactly how the measurement system will follow these variations: the dynamic behaviour must permit a true reproduction of the measurand.

The dynamic behaviour of a *linear measurement system* is fully determined by its response to a single step function at the input (*step response*). From the step response we can see the *settling time* or the *reading time* of the measurement system. Fig. 2.36 shows both the step variation at the input $x(t)$ and the response $y'(t)$ of the measurement system. The response has been normalised here with respect to the DC sensitivity $S(0)$, so the limit for $t \to \infty$ gives coinciding values of input and output. The settling time is the time from the occurrence of the step at the input to the point in time after which the output signal $y(t)$ remains for the first time within a specified tolerance interval ($y_o - \Delta y_o$, $y_o + \Delta y_o$) around the end value y_o. The ratio $\pm \Delta y_o / y_o$ is the relative inaccuracy of the measurement system. The settling time is a measure for the maximum measuring rate of the system.

The dynamic behaviour of a linear measurement system is also fully defined when we know how the system responds to a sine wave of varying frequency; i.e. when we know the *frequency response*.

Determining the frequency response of a measurement system results in the (complex) sensitivity $\overline{S}(\omega)$. The system sensitivity $\overline{S}(\omega)$ has two components: the *amplitude response* $|\overline{S}(\omega)|$ and the *phase response* Arg $\overline{S}(\omega)$. Together, these two define the dynamic behaviour of a linear system completely. A measure of the frequency capability is the *bandwidth f_0*. This is the frequency at which the power of the output signal has dropped to half the maximum output power. Thus, at f_0, the amplitude of the output signal has decreased to $1/\sqrt{2}$ of the low-frequency value. The amplitude response at this frequency f_0 is therefore only $S(0)/\sqrt{2}$ compared to the DC-value of $S(0)$. This is illustrated in Fig. 2.37. Since $20\log_{10}(1/\sqrt{2}) \approx -3$, this frequency f_0 is also called the *minus 3 decibel frequency* (abbreviated to –3 dB frequency). The (deci)bel is a logarithmic measure for a power ratio. This measure is discussed in Section 5.3.

The relationship between the input quantity $x(t)$ and the output quantity $y(t)$, and therefore the dynamic behaviour of a *linear* dynamic system, can also be given in the form of a *linear* differential equation.

Figure 2.36. The response $y(t)$ of a linear measurement system to a step function $x(t)$ on the input. The step response has been normalised, so $y'(t) = y(t)/S(0)$ where $S(0)$ is the DC sensitivity of the system. The instrument settling time for an inaccuracy of $\pm \Delta y_0/y_0$ is equal to t_s.

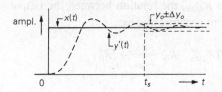

Figure 2.37. Frequency response $\overline{S}(f)$, consisting of the amplitude characteristic $|\overline{S}(f)|$ and the phase characteristic Arg $\overline{S}(f)$. The system bandwidth is f_0.

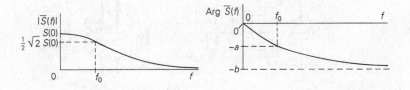

If y depends on x, then a differential equation of y and x contains not only functions of y and x, but also the time derivatives of these functions. A linear differential equation will contain only first order terms in y and first order derivatives. The order of a differential equation is equal to the order of the highest derivative.

Many different measurement systems (electrical, thermal, acoustic, etc.) exhibit a similar dynamic behaviour. If we generalise the description of these systems by means of across and through quantities, we will find the same differential equation. Hence, in our exploration of dynamic systems, we may limit ourselves to a restricted number of differential equations. The order of a differential equation is particularly important for the dynamic behaviour. We therefore speak of an nth *order (measurement) system*, with an nth order response, if the behaviour of the system can be expressed in the form of an nth order differential equation. In practice, we can usually describe most measurement systems sufficiently accurately by means of a second or lower order linear differential equation. Hence, a discussion of the values 0, 1 and 2 for n will suffice.

Zero order systems
The differential equation which describes a zero order system is no more than a simple algebraic ordinary equation. The system is static or, in other words, independent of frequency. An example of a zero order system is the potentiometric displacement gauge, depicted in Fig. 2.38. This transducer converts a displacement x into a proportional output voltage V. Assuming the resistance between the bottom connection of the potentiometer and the wiper equals R_0 when x is zero, and that for $x = x_{max}$ the maximum resistance equals R_{max}, the relation between the output voltage V and the displacement x can be expressed as:

$$V = I(R_{max} - R_0) \frac{x}{x_{max}} + IR_0$$

Figure 2.38. Zero order systems: (a) displacement potentiometer and (b) current probe.

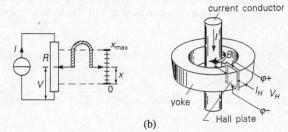

(a)   (b)

The form of this equation is the same as:

$$y = ax + b$$

Here b is the zero or offset value and a is the sensitivity. The settling time t_s is zero and the bandwidth f_0 is infinite. In reality of course, the sensitivity will decrease for very high frequencies due to many causes (elasticity, mass, parasitic capacitance, etc.), and therefore we often speak of *quasi-zero order systems*. This means that these systems respond instantaneously over the frequency range which is of interest for measuring the input quantity. The system of Fig. 2.38(b) can also be considered to be a zero order system. The output voltage V_H of the Hall device is proportional to the current I_H flowing through the Hall plate and the magnetic induction B which is generated by the central conductor. In addition, there will also be an offset voltage V_o. In Section 3.2.3 we will show that:

$$V_H = R_H \frac{I_H B}{t} + V_o$$

Here R_H is the Hall coefficient of the Hall plate and t is its thickness. Since B is proportional to the current flowing through the central conductor, this transducer is extremely suitable for measuring large currents. It can do so without the need to open up the conductor and insert a current probe which would introduce insertion resistance.

First order systems
The relation between the input signal $x(t)$ and the output signal $y(t)$ for a linear first order system is determined by a linear first order differential equation. A simple example of such a system is the mercury thermometer. The expansion of the mercury column along the calibrated scale is the output signal and the measured ambient temperature T_i is the input signal. We will assume that the change in length of the mercury column is linearly proportional to the change in temperature of the mercury in the reservoir of the thermometer. Therefore, to describe the dynamic behaviour of the thermometer, we can just as easily take the temperature T_o of the mercury in the reservoir as the output quantity (see Fig. 2.39(a)). The heat exchange between the reservoir and the surrounding air must occur through a glass wall with a thermal resistance R. The thermal capacitance (heat capacity) of the mercury in the reservoir is denoted by C. For a small increase in heat, ΔQ, we find:

$$\Delta Q = C \, \Delta T_o$$

In addition:

$$\Delta Q = I_h \, \Delta t = \frac{T_i - T_o}{R} \Delta t$$

and therefore:

$$RC \frac{dT_o}{dt} + T_o = T_i$$

Another example of a first order system is the *RC* network in Fig. 2.39(b). The input voltage is V_i and the output voltage is V_o. With:

$$V_o + IR = V_i$$

and:

$$I = \frac{dQ}{dt} = C \frac{dV_o}{dt}$$

we get:

$$RC \frac{dV_o}{dt} + V_o = V_i$$

With $RC = \tau$ we can generalise these two linear first order differential equations into the following form:

$$\tau \frac{dy}{dt} + y = x$$

where $x = x(t)$ and $y = y(t)$. We, therefore, may conclude that, as far as their dynamic behaviour is concerned, the mercury thermometer and the RC network are equivalent.

Solving the differential equation is simple. With a step input:

$$x(t) = x_0 \quad \text{when } t \geq 0$$

and

Figure 2.39. Two examples of first order systems: (a) a mercury thermometer and (b) an *RC* circuit.

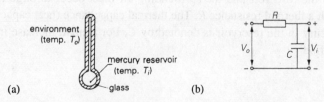

environment
(temp. T_o)

mercury reservoir
(temp. T_i)

glass

(a)   (b)

$$x(t) = 0 \quad \text{when } t < 0$$

the output $y(t)$, or the step response, is found from:

$$\tau \frac{dy}{dt} + y = x$$

The complementary equation is given by:

$$\tau p + 1 = 0$$

thus:

$$p = -1/\tau$$

The general solution is: $y = C\,e^{-t/\tau}$

A particular solution when $t \to \infty$ is $y(t) = y_0$ (the final or end value):

$$y = C\,e^{-t/\tau} + y_0$$

Assuming that when $t = 0$, $y(t) = 0$, we find:

$$y(t) = y_0(1 - e^{-t/\tau})$$

The *step response* of a first order measurement system is given by:

$$y = y_0(1 - e^{-t/\tau})$$

in which y_0 is the final or end value and τ the *RC* time. This step response is illustrated in Fig. 2.40.

If the relative inaccuracy of a measurement system may not exceed $\varepsilon = \Delta y_0 / y_0$, then the settling time t_s is given by:

$$t_s = -\tau \ln \varepsilon$$

Figure 2.40. Step response of a first order system.

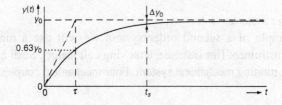

This can easily be derived from the expression for the step response with the aid of Fig. 2.40. When a very small relative read out error $\Delta y_o/y_o$ is required in such a first order system, the settling time t_s will become excessively large. In the example of the thermometer of Fig. 2.39(a) the mercury column increasingly slowly creeps up to the final length which is attained when the system is in thermal equilibrium with its environment. Therefore, a first order response is undesirable if an accurate measurement is required; the settling time will be too long.

The *frequency response* is, in fact, the response of a system to a sinusoidal input signal $x(t) = \hat{x} \sin(\omega t)$ in the *steady state*: i.e. the response at a time, long after switching on and applying the input signal; at a time when all transient phenomena have died out. The frequency response follows from the particular solution for $t \to \infty$ of the differential equation describing the first order system.

One can easily compute the frequency response using complex variables in the electrical analogue of the first order system given in Fig. 2.39(b). This results in:

$$\overline{S}(\omega) = \frac{1}{1 + j\omega\tau} \quad \text{(with } \tau = RC\text{)}$$

The magnitude of $\overline{S}(\omega)$ is given by:

$$|\overline{S}(\omega)| = \frac{1}{\sqrt{1 + \omega^2\tau^2}}$$

and the argument of $\overline{S}(\omega)$ is

$$\text{Arg } \overline{S}(\omega) = -\arctan \omega\tau$$

The amplitude characteristic $|\overline{S}(\omega)|$ and the phase characteristic $\text{Arg } \overline{S}(\omega)$ of a first order system are plotted in Fig. 2.41.

The bandwidth f_0 of a first order system is defined by $\omega\tau = 1$, so $f_0 = 1/2\pi\tau$. The phase shift at this frequency already amounts to $-\pi/4$ or $-45°$. For higher frequencies the phase shift approaches $-90°$, while the amplitude of the output signal reduces to nearly zero.

Second order systems
As an example of a second order system we will use a mechanically indicating instrument (for instance, a moving coil meter). Such a meter will consist of a rotating mechanical system. Four mechanical couples each exert

a torque on the rotating part of the meter, as shown in Fig. 2.42(a). These are:

- *Deflecting torque*. This torque causes the pointer to deflect over an angle θ. Its moment is proportional to the measured quantity (current). We denote this moment M_a.
- *Restoring torque*. The moment associated with this torque opposes the deflection. It is provided here by a spiral spring and is denoted M_r. Once the steady state is reached, the deflecting moment and the restoring moment are equal: $M_d = M_r$. Usually the restoring moment is made proportional to the deflection angle θ, so $M_r = K_r\theta$, in which K_r is the spring constant (spring stiffness).
- *Damping torque*. The moment of this torque also opposes the deflecting moment. The damping moment is proportional to the angular velocity of the pointer, so $M_{da} = D_r\, d\theta/dt$. Here D_r is the damping constant of the rotating part. The damping increases linearly with the angular velocity $d\theta/dt$.

Figure 2.41. Frequency response of a first order system.

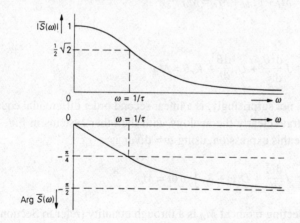

Figure 2.42. Second order systems: (a) mechanical rotation; (b) mechanical translation; (c) electrical parallel network.

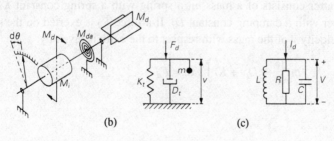

(a)   (b)   (c)

Damping is used to prevent the pointer overshooting the final value and ringing out around the final value. Methods for realising damping include the use of vanes, or pistons (air damping), or the induction of eddy currents in a metal plate when the system is moving (Foucault current damping).

- *Inertial torque*. The inertia of the rotating part of the meter gives rise to another counteracting torque, the moment of which is proportional to the angular acceleration of the pointer, so:

$$M_i = J \frac{d^2\theta}{dt^2}$$

where J is the moment of inertia of the rotating part around the axis of rotation.

The dynamic behaviour of the meter is determined by the *equation of motion* of the meter; the deflecting moment will constantly be balanced by the sum of all other moments:

$$M_i + M_{da} + M_r = M_d$$

Or:

$$J \frac{d^2\theta}{dt^2} + D_r \frac{d\theta}{dt} + K_r\theta = M_d$$

The result, not surprisingly, is a linear second order differential equation.

To illustrate clearly the analogy with the other systems in Fig. 2.42, we can rewrite this expression, using $\omega = d\theta/dt$ as:

$$J \frac{d\omega}{dt} + D_r\omega + K_r \int \omega \, dt = M_d$$

The deflecting moment M_d is a through quantity (refer to Section 5.4) and the angular velocity ω an across quantity. This rotating mechanical system is analogous to the system of Fig. 2.42(b), which is a translational system. The latter consists of a mass m, a spring with a spring constant K_t and a damper with a damping constant D_t. If a force F_d is exerted on the system, the velocity v of the mass with respect to the earth is given by:

$$m \frac{dv}{dt} + D_t v + K_t \int v \, dt = F_d$$

Since $v = dx/dt$, this is again the same linear second order differential equation that we encountered above. Furthermore, the mechanical rotating and translating systems are analogues of the electrical system in Fig. 2.42(c). This electrical parallel network is driven by a through quantity, the drive current I_d. We wish to determine the across quantity V, which is given by:

$$C\frac{dV}{dt} + \frac{V}{R} + \frac{1}{L}\int V\,dt = I_d$$

Again, this equation is equivalent to the two former equations. All we did was exchange the through quantities and the across quantities for one another. The structure of the networks remains the same if we interchange J, m and C, as well as D_r, D_t and $1/R$ and also K_r, K_t and $1/L$ (see Section A.4). Using:

$$V = L\frac{dI}{dt}$$

the last expression can be rewritten as:

$$LC\frac{d^2I}{dt^2} + \frac{L}{R}\frac{dI}{dt} + I = I_d$$

The current I here represents the current through the coil L.

We have now established that the general form of the differential equation which describes linear second order systems contains two constants, a and b:

$$a\frac{d^2y}{dt^2} + b\frac{dy}{dt} + y = x$$

Here, x is the driving input quantity $x(t)$ and y is the output quantity $y(t)$ which has been normalised with respect to the DC sensitivity $S(0)$, so $y = y(t)/S(0)$. This causes the third constant in the differential equation to disappear. In order to provide a clearer expression, we can introduce two other constants, the relative damping z and the angular frequency ω_0 of the free, undamped resonance of the system, and rewrite the general equation above as:

$$\frac{1}{\omega_0^2}\frac{d^2y}{dt^2} + \frac{2z}{\omega_0}\frac{dy}{dt} + y = x$$

Then, for a rotational system, the variables and parameters become:

$$x = \frac{M_d}{K_r}, \qquad y = \theta, \qquad \omega_0^2 = \frac{K_r}{J} \qquad \text{and} \qquad z^2 = \frac{D_r^2}{4K_rJ}$$

and for a translational system:

$$x = \frac{F_d}{K_t}, \qquad y = x, \qquad \omega_0^2 = \frac{K_t}{m} \qquad \text{and} \qquad z^2 = \frac{D_t^2}{4K_tm}$$

For the electrical network we find:

$$x = I_d, \qquad y = I, \qquad \omega_0^2 = \frac{1}{LC} \qquad \text{and} \qquad z^2 = \frac{L}{4R^2C}$$

The corresponding complementary equation is:

$$\frac{1}{\omega_0^2} p^2 + \frac{2z}{\omega_0} p + 1 = 0.$$

the roots of which are given by:

$$p_{1,2} = -\omega_0 z \pm \omega_0 \sqrt{z^2 - 1}$$

We must distinguish three characteristic cases: $z < 1$, $z = 1$ and $z > 1$.

Underdamping ($z < 1$)
It can be shown that the solution for the system response $y(t)$ to a step input
signal at $t = 0$ of magnitude x_0 is given by:

$$y(t) = x_0 \left[1 - \{\cos(\omega_i t) + \frac{z\omega_0}{\omega_i} \sin(\omega_i t)\} \exp(-z\omega_0 t) \right]$$

where $\omega_i = \omega_0 \sqrt{1 - z^2}$ and the assumption is made that the initial values $y(0)$
and $(dy/dt)_{t=0}$ are both zero. The final value or steady state value is given
by:

$$y_0 = \lim_{t \to \infty} y(t) = x_0$$

Immediately after the input is applied, a damped oscillation with a frequency ω_i is superimposed on the final value (see Fig. 2.43). We observe
that when z increases, the damping of the ringing, increases transiently.
Therefore, z is called the *relative damping ratio*. When z is zero, the system
will continue to oscillate around the final value at a frequency ω_0; the
system will exhibit *free vibrations*. Thus, the frequency ω_0 is the resonance
frequency of the system without any damping at all.

Critical damping (z = 1)

Assuming again the initial conditions $y(0) = 0$ and $(dy/dt)_{t=0} = 0$, and that the magnitude of the input step at $t = 0$ equals x_0, the step response of a second order system with $z = 1$ is given by:

$$y(t) = x_0\{1 - (1 + \omega_0 t)\, e^{-\omega_0 t}\}$$

Again, the final value is $y_0 = x_0$, but now, the output exhibits no damped ringing (see Fig. 2.43).

Usually, moving coil meters are designed to be just a little underdamped ($z = 1/\sqrt{2}$), rather than exactly critically damped. This causes the pointer to overshoot slightly (4%). This has the advantage that the observer can clearly see when the pointer has reached its final value. Another advantage of choosing $z \approx 0.7$ in measurement systems is that for this value the amplitude characteristic is flat over the widest possible frequency range (this will be discussed in Fig. 2.45).

Overdamping (z > 1)

With the same initial conditions as above, but with a damping z larger than one, the step response $y(t)$ to an input step x_0 at $t = 0$ is given by:

$$y(t) = x_0\left[1 - \{\cosh(\omega_i t) + \frac{z\omega_0}{\omega_i} \sinh(\omega_i t)\}\, e^{-z\omega_0 t} \right]$$

where $\omega_i = \omega_0\sqrt{z^2 - 1}$. The output will now 'creep' to the final value $y_0 = x_0$ (see Fig. 2.43).

Figure 2.43. The step response of a second order system for various values of the relative damping z.

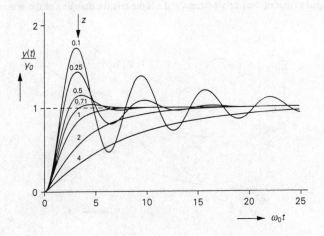

The instrument or settling reading time t_s of a second order system depends on the damping z, the period T_0 which corresponds to the frequency of the free vibrations ($T_0 = 2\pi/\omega_0$) and, of course, the allowed relative error in the final value $\Delta y_0/y_0$ (see Fig. 2.44). The curves in this figure exhibit discontinuities for $z < 1$, which are caused by the fact that the reading time t_s (for a given constant value of T_0) increases in discrete steps for a given relative inaccuracy (say 0.1%) and a continuously decreasing damping z. The reason is that the reading time will increase each time by one period of the ringing transient.

The frequency response of a second order system can easily be calculated with the *RLC* analogue of the second order system shown in Fig. 2.42(c):

$$\overline{S}(\omega) = \frac{\overline{y}(\omega)}{\overline{x}(\omega)} = \frac{1}{1 + j\omega L/R - \omega^2 LC}$$

The amplitude characteristic follows from:

$$|\overline{S}(\omega)| = \frac{1}{\sqrt{\omega^2 L^2/R^2 + (1 - \omega^2 LC)^2}}$$

and the phase characteristic is given by:

$$\text{Arg } \overline{S}(\omega) = -\arctan\left\{\frac{\omega L}{R(1 - \omega^2 LC)}\right\}$$

Figure 2.44. The reading or settling time t_s of a second order system for different values of the allowed relative error $\Delta y_0/y_0$ in the final value y_0. The period T_0 is equal to that of the free vibrations and z is the relative damping of the system.

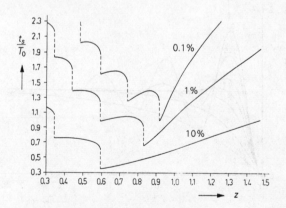

These expressions can be written in a general form by substituting $z^2 = L/4R^2C$ and $\omega_0^2 = 1/LC$. Differentiating $|\overline{S}(\omega)|$ with respect to ω shows that the maximum of $|\overline{S}(\omega)|$ occurs when $\omega_{max} = \omega_0\sqrt{1 - 2z^2}$ and the value of this maximum is given by:

$$|\overline{S}(\omega_{max})| = \frac{1}{2z\sqrt{1 - z^2}} \quad (\text{when } z \leq \tfrac{1}{2}\sqrt{2})$$

The frequency response is flat over the widest possible frequency range if $|\overline{S}(\omega_{max})| = |\overline{S}(0)| = 1$, which is the case if $z = \tfrac{1}{2}\sqrt{2}$. Then, the bandwidth of the system is $f_0 = \omega_0/2\pi$, with ω_0 the frequency of the free vibrations. The phase shift at $\omega = \omega_0$ is equal to $-90°$ and at very high frequencies the phase shift will approach but never exceed $-180°$. The output signal of the system is then nearly zero. If $z < \tfrac{1}{2}\sqrt{2}$, the amplitude characteristic will exhibit a peak at the damped resonance frequency (see Fig. 2.45(a) and (b)).

Figure 2.45. (a) Amplitude and (b) phase characteristics of a second order system for various values of the damping z.

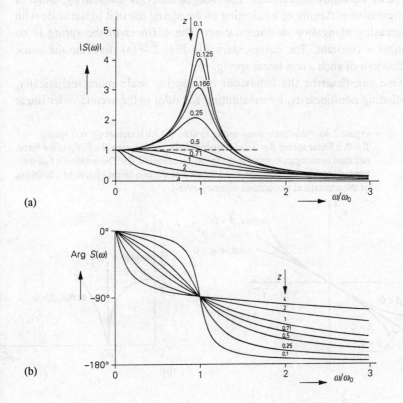

(a)

(b)

Non-linear systems

So far, we have assumed that the response of the systems we were studying is linear. But what happens if these systems are *non-linear*? In Fig. 2.32 it has already been shown that, provided the input quantity becomes large enough, every practical system will eventually become non-linear, due to saturation, overloading or clipping. We will now illustrate what happens to the response in such a case, based on a simple mechanical example.

Consider the classical method of measuring a mechanical force by means of a spring scale which converts the force F_d applied to the internal spring into a proportional elongation l of the spring. The pin or ring which couples the force into the spring will have a certain mass and the spring is thought to have a certain damping. Thus, for a small input force and a resulting small elongation we can use the analogue shown in Fig. 2.42(b). If l is small, the static behaviour of the spring scale is governed by Hooke's law: $F_d = K_t l$, in which K_t is the spring stiffness (spring constant). The dynamic behaviour of this linear system has already been discussed; the spring scale for small loads is a second order, linear damped mass-spring system.

When a large force is applied, however, this no longer holds true; the system becomes non-linear. The non-linearity is caused by either a progressive stiffening or weakening of the spring internal to the scale with increasing elongation or compression. The stiffness of the spring is no longer a constant. The curves shown in Fig. 2.46(a) illustrate the static behaviour of such a non-linear spring.

One can describe the behaviour of a spring scale more realistically, including non-linearity, by substituting $v = dl/dt$ in the second order linear

Figure 2.46. Non-linear mass-spring system. (a) An increasingly stiff spring $\beta > 0$, a linear spring $\beta = 0$ and an increasingly weak spring $\beta < 0$; F_d is the force required to elongate/compress the spring an amount l. (b) The amplitude l of the fundamental harmonic of the free vibration of the non-linear system as a function of the normalised resonance frequency ω/ω_0.

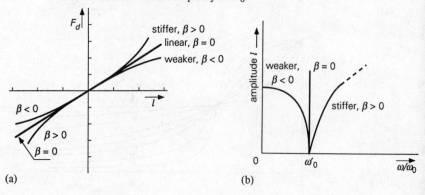

(a)   (b)

differential equation associated with a mechanical translational system. The result is:

$$m \frac{d^2 l}{dt^2} + D_t \frac{dl}{dt} + K_t l + \beta l^3 = F_d(t), \quad \text{and} \quad l = l(t)$$

The non-linearity originates from the fourth term in the left-hand side of the first equation. The non-linearity assumes the spring to be symmetrical (the same non-linearity for elongation as for compression). The non-linearity arises from the presence of l to a power different from 1. If $\beta > 0$ the spring will become increasingly stiffer the further it is elongated or compressed. If $\beta = 0$ the spring is linear, and if $\beta < 0$ the spring will become increasingly less resilient with l. For very small values of l, the system is reduced to a linear, second order system, as for this case $\beta l^3 \ll K_t l$ and therefore the term βl^3 causing the non-linearity may be neglected.

In order to establish the exact dynamic behaviour we will first consider the free vibrations of the system: $F_d(t) = 0$ for $t \geq 0$. In addition, we will also assume an undamped system: $D_t = 0$. Once the system is mechanically excited, it will continue to oscillate producing a periodic signal $l = l(t) = l(t \pm nT)$, n integer. The period $T = 1/f_0 = 2\pi/\omega$ now becomes a function of the *amplitude* of the vibration! This is indicated in Fig. 2.46(b). For small excursions ω becomes equal to the angular frequency of the *linear* second order system ω_0. As the amplitude of the excursions increases, the period will become shorter or longer, depending on whether the spring becomes progressively stiffer or weaker ($\beta > 0$ or $\beta < 0$). It should be noticed that now the waveform of $l = l(t)$ is no longer a pure sine wave, but contains higher harmonics in addition to the fundamental. The *waveform* will also change as the amplitude of the vibration increases. The waveform is only sinusoidal for very small excursions.

The above discussion will have made it clear that for non-linear systems one can no longer speak of the *frequency response*, since the behaviour has become dependent on the amplitude! The principle of superposition no longer holds, and we will see in the following that these systems may exhibit an extraordinary dynamic behaviour.

Assuming now that the damping is no longer equal to zero ($D_t \neq 0$), we will apply a sinusoidal external driving force to the system $F_d = F_d(t)$, with a constant amplitude $\hat{F}_d$. Fig. 2.47 shows what happens to such a system with a spring that progressively stiffens ($\beta > 0$). Here the amplitude $l = l(t)$ of the fundamental frequency has been plotted against the forcing frequency ω.

To obtain this curve, the fundamental harmonic frequency which coincides with the (forcing) frequency of the driving force has been filtered out from the (distorted) signal $l(t)$. Here, only the amplitude of this fundamental harmonic is plotted (linearised behaviour). As we see, the resonance curve no longer has the form shown in Fig. 2.45(a) for a *linear* second order system. It now leans towards higher frequencies: in other words, it is skewed. If the amplitude $\hat{F}_d$ of the sinusoidal driving force is varied, the peak of the resonance curve will follow the dashed curve in Fig. 2.47 which describes the resonance of the free vibrations (as in Fig. 2.46(b), $\beta > 0$).

If the frequency of the driving force is increased, but the amplitude $\hat{F}_d$ is kept constant, the output of the mass-spring system will suddenly drop to a much lower level (*jumping phenomenon*) after reaching the peak of the resonance curve at $\omega = b\omega_0$. When the frequency is decreased (again, with a constant input force amplitude), the amplitude of the fundamental harmonic in $l(t)$ will suddenly jump to a higher value at the frequency $\omega = a\omega_0$. Consequently, in the frequency interval $a\omega_0 \leq \omega \leq b\omega_0$, the system is *unstable*. The steady state of the system can never lie on the dotted part of the resonance curve within this interval. We see that the resonance curves of this kind of non-linear mechanical system exhibit *hysteresis*.

If, in addition to this, we now also vary the *amplitude* of the driving force $F_d(t)$, we will observe the resonance curves of Fig. 2.48. Fig. 2.48(a) shows the curves of a mass-spring system with a spring of progressively increasing stiffness, while Fig. 2.48(b) shows the curves for a spring with decreasing stiffness.

Figure 2.47. Resonance curve of the amplitude of the fundamental harmonic of a non-linear damped mass-spring system (spring with increasing stiffness) for a forced vibration resulting from a sinusoidal driving force with constant amplitude.

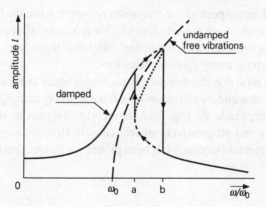

Finally, it is observed that when this kind of system is excited with a sine wave force (of frequency ω) it can exhibit *subharmonics* at ω/n besides *higher harmonics* at $n\omega$ (where n is an integer). This usually occurs when the damping D_t is made small, but non-zero.

This illustration of the dynamic behaviour of a non-linear dynamic system clearly shows how complex these systems can be. This is one of the reasons why one avoids any essential (dynamic) non-linearity in measurement systems; the system becomes too complex.

2.3.3.3 Disturbances

In Section 2.3.3 we have seen that the interaction between a measurement system and a measurement object not only consists of the desired influence of the object on the measurement system, but also comprises an undesired reaction of the object to the presence of measurement system. It has been discussed that the latter reaction can be reduced considerably by matching the measurement system to the measurement object.

Figure 2.48. The influence of the amplitude of the forcing input function on the resonance curves (see Fig. 2.47) of a non-linear, damped mass-spring system. In (a) the case is shown for a spring of increasing stiffness ($\beta > 0$) and in (b) the amplitud—frequency curves of a progressively softening spring ($\beta < 0$) are shown.

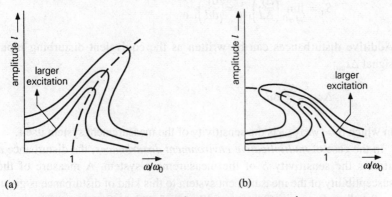

(a)   (b)

Figure 2.49. Interaction between the measurement system, the measurement object, the observer and the surroundings (x is the input signal, y is the output signal and d is a disturbance).

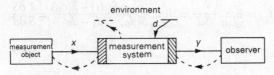

In this section we will examine the interaction between the measurement system and its environment (see Fig. 2.49). The influence of the environment on the measurement system can affect the measurement result and is therefore an undesirable and disturbing influence.

Such a disturbing environment can influence a measurement system in two ways: additively or multiplicatively. In the case of an *additive disturbance* the disturbing environment parameter d will cause a measurement system output signal y, even when there is no input signal x (no measurand applied). This output is added to the output caused by a measured input x. The system responds as if a transfer function existed by which the disturbance d could 'leak' into the output y. Therefore, the measure of the extent to which the output y is affected by the disturbance d is given in the form of a sensitivity: the *disturbance sensitivity* (or sensitivity factor) S_d:

$$S_d = \left(\frac{y}{d}\right)_{x=0}$$

Since this transfer function can be non-linear and the actual disturbance is often in the form of a *variation* Δd in an environment parameter d (as is the case with, for instance, the ambient temperature or the supply voltage of a system), a more correct notation is:

$$S_d = \lim_{\Delta d \to 0} \left(\frac{\Delta y}{\Delta d}\right)_{x=0} = \left(\frac{dy}{dd}\right)_{x=0}$$

Additive disturbances can be written as the equivalent disturbing input signal Δx_{eq}:

$$\Delta x_{eq} = \frac{S_d}{S} \Delta d$$

in which $S = (y/x)_{\Delta d=0}$ is the sensitivity of the measurement system itself.

In the case of *multiplicative environment disturbances*, the disturbance d affects the sensitivity S of the measurement system. A measure of the susceptibility of the measurement system to this kind of disturbance is given by the *disturbance coefficient* (or influence function) C_d:

$$C_d = \lim_{\Delta d \to 0} \frac{\left(\dfrac{\Delta S}{S}\right)}{\Delta d} = \lim_{\Delta d \to 0} \frac{1}{S(d_0)} \frac{S(d_1) - S(d_0)}{d_1 - d_0} = \frac{1}{S}\frac{\partial S}{\partial d}$$

with $d_1 - d_0 = \Delta d$.

Nomenclature: When dealing with a specific disturbance, the name of this specific disturbance is often used to describe the measures we have just introduced. For instance, we speak of the *supply voltage sensitivity S_v* and the *temperature coefficient C_T* of a measurement system.

Even when there is no input voltage applied to a DC-voltage null detector, the pointer will indicate a small deflection. The magnitude of this deflection will depend partially on the supply voltage. The supply voltage sensitivity S_v to this *additive disturbance* may, for example, therefore be 10 μV/V = 10^{-5} (dimensionless).

The sensitivity (gain) of an instrumentation amplifier is determined by the ratio of two resistors, whose values depend (to a certain extent) on the ambient temperature. For a *multiplicative disturbance* like this the temperature coefficient C_T may be stated as, 10^{-5}/K, for instance.

The influence of environmental disturbances on a measurement system can be kept to a minimum by the choice of the least sensitive measurement system. In addition, it is also possible to take one or more of the following steps to reduce this unwanted environmental influence even further:

- *Isolate* the measurement system (or parts thereof) from external disturbances: for example, electrical shielding, stabilisation of the ambient temperature, keeping the humidity constant, etc.
- Use a measurement method with a *low intrinsic disturbance sensitivity S_d*. For example, we can first measure the output signal y_0 due to the additive disturbance only ($x = 0$), then measure the output y due to the input signal plus the disturbance, and finally, measure again the output y_1 due to only the disturbance. The disturbance corrected measurement result will then be $y - \frac{1}{2}(y_0 + y_1)$.
- Build measurement systems with components which have a *low disturbance coefficient C_d*: for example, use components with a small temperature coefficient.
- Design the measurement system in such a way that, otherwise unavoidable disturbances *compensate* each other through parallel, series or ratio compensation. These three methods of compensation are illustrated in Fig. 2.50.

With *parallel compensation* the measurement system (or the critical part thereof) is split into two parallel parts S_1 and S_2, which are both exposed to the same external disturbance d. The sensitivity of the system S is made equal to $S = S_1 + S_2$ and is no longer affected by any *additive disturbance* if the respective disturbance sensitivities of subsystem S_1 and S_2 are:

$$S_{d1} = -S_{d2}$$

whereas S will not be sensitive to *multiplicative disturbances* if:

$$S_1 C_{d1} = -S_2 C_{d2}$$

For the method of *series compensation* the measurement system is split into two parts, which are connected in series and which are both exposed to the same external disturbance d. The sensitivity of the system is $S = S_1 S_2$. If:

$$S_2 C_{d1} = -S_{d2}$$

the system will no longer be affected by *additive disturbances*. The effect of *multiplicative disturbances* is fully cancelled when:

$$C_{d1} = -C_{d2}$$

It should be noted that the last expression only holds true when the disturbance d in S_1 and S_2 causes relative changes in the subsystem sensitivities S_1 and S_2 (respectively denoted $\Delta S_1 / S_1$ and $\Delta S_2 / S_2$) that are much smaller than 1.

Finally, the method of *ratio compensation* is only effective in combating *multiplicative disturbances*. The sensitivity of a ratio system is equal to $S = S_1 / S_2 x_{\text{ref}}$ in which x_{ref} is a constant reference quantity. A multiplicative external disturbance d affects both S_1 and S_2. This multiplicative disturbance cancels in the output if:

$$C_{d1} = C_{d2}$$

An example of parallel compensation is a long-tailed pair differential amplifier. The zero-offset error of 650 mV, with a temperature coefficient of 2.5 mV/K of the two bipolar input transistors, is reduced here to approximately 1 mV with a temperature coefficient ± 2 μV/K. Series compensation is often applied to disturbances in measurement transducers. The distur-

Figure 2.50. The three basic methods for compensating environmental disturbances: (a) parallel compensation, (b) serial compensation and (c) quotient compensation.

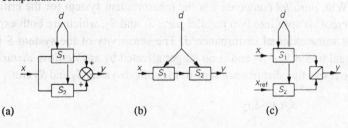

(a)   (b)   (c)

bance is compensated in a network placed in series with the measurand. Finally, ratio compensation occurs *de facto* in every ratio measurement.

- *Change* the wave form or the frequency spectrum of the *input signal* in such a way that the disturbance can easily be separated from the useful signal. For instance, an indirect DC-voltage amplifier (see Section 3.3.4) converts the applied DC input voltage to an alternating voltage. This avoids the effects of DC offset and drift within the amplifier itself.
- Use *feedback* against multiplicative disturbances. In Fig. 2.51 feedback is implemented by subtracting a fraction β of the output signal y from the input signal x of a measurement system which has a sensitivity S_0. The sensitivity of the overall system with feedback S_f is no longer equal to S_0, but is given by:

$$S_f = \frac{S_0}{1 + \beta S_0}$$

Suppose S_0 depends on the temperature and has a temperature coefficient:

$$C_{T0} = \frac{(\Delta S_0/S_0)}{\Delta T}$$

The temperature coefficient C_{Tf} of the system with feedback will then be:

$$C_{Tf} = \frac{(\Delta S_f/S_f)}{\Delta T}$$

Since:

$$S_f + \Delta S_f = \frac{S_0 + \Delta S_0}{1 + \beta(S_0 + \Delta S_0)}$$

it follows that:

$$\frac{\Delta S_f}{S_f} = \frac{(\Delta S_0/S_0)}{1 + \beta S_0(1 + \Delta S_0/S_0)}$$

For small variations $\Delta S_0/S_0 \ll 1$:

Figure 2.51. Feedback applied to a measurement system S_0.

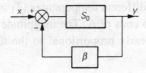

$$\frac{\Delta S_f}{S_f} \approx \frac{(\Delta S_0/S_0)}{1 + \beta S_0}$$

and, therefore, also:

$$C_{Tf} \approx \frac{C_{T0}}{1 + \beta S_0}$$

We refer to the case $|1 + \beta S_0| > 1$ as *negative feedback*, whereas for $|1 + \beta S_0| < 1$ we speak of *positive feedback*. When the negative feedback is large, the disturbance coefficient will be reduced considerably. However, for a large negative feedback the sensitivity, which is then $S_f \approx 1/\beta$, is far less than the original system sensitivity S_0. Therefore, for an accurately defined sensitivity S_f, free of disturbances, the transfer function β must be as accurate and free of disturbances as possible. β now determines the system sensitivity. The penalty we pay is a reduction in system sensitivity from the original value of S_0 to $1/\beta$.

Note that negative feedback reduces *additive disturbances* by the same factor as that by which the sensitivity of the system is reduced. This means that the ratio of the measurement signal and the disturbances (both referred to the output or to the input) *will not change* due to the application of feedback. In the same way, the signal-to-noise ratio of the measurement system will also not be improved by using negative feedback. We will return to the application of negative feedback in Section 3.3.4.

Let us now examine several frequently encountered sources of disturbance. We will confine ourselves to sources which occur in *electronic* measurement systems.

Thermoelectricity

Thermoelectricity arises when a conductor circuit in a measurement system is composed of two or more dissimilar materials (or more precisely: dissimilar materials or materials in *different states*, e.g. one under strain and the other not). This can be observed in the form of a thermally induced potential difference (contact potential difference) across the junction of these materials when they are at different temperatures. The potential difference depends on the temperature drop across the junction. Therefore, if the two junctions between sections of different metallic conductors A and B in Fig. 2.52 are held at different temperatures ($T_1 \neq T_2$), a voltage will be measurable across the two ends of A. The magnitude of this thermoelectric voltage V is (almost) linearly proportional to the temperature difference $T_1 - T_2$.

The thermoelectric voltage also depends on the type of metals used in the conductors:

$$\left.\begin{array}{l} \text{Cu–Ag} \\ \text{Cu–Au} \\ \text{Cu–Cd/Sn} \end{array}\right\} 0.3\ \mu\text{V/K} \qquad \begin{array}{ll} \text{Cu–Pb/Sn:} & 3\ \mu\text{V/K} \\ \text{Cu–Kovar:} & 500\ \mu\text{V/K} \\ \text{Cu–CuO:} & 1000\ \mu\text{V/K} \end{array}$$

The materials in this list are:
- Pb/Sn: ordinary solder;
- Cd/Sn: special solder for junctions with small thermoelectric voltages;
- Kovar: material used for the pins of semiconductors;
- CuO: copper oxide. (The surface of a connector is covered with an oxide layer. This layer must be broken by screwing the connector down tightly.)

Thermoelectric voltages in a measurement system will cause *zero-offset errors* and are therefore a form of *additive disturbance*. In order to avoid thermal voltages the conductors within a measurement system and those attached externally to the input must be of the same material. If this is not possible, then combinations of metals should be used which produce a low thermal voltage. Also, the system must be constructed in such a way that a large temperature difference between the input connector is avoided.

Leakage currents

If two conductors at different electrical potentials are not perfectly insulated, an unwanted current will flow between the two conductors. The magnitude of this current depends on the insulation material, the level of impurities in the insulation and the humidity of the surrounding air, accumulated debris and surface films, etc. Circuits at a high impedance level (a high source impedance as well as a high system input impedance)

Figure 2.52. Thermoelectricity as an additive disturbance in electrical systems.
(a) Elementary thermocouple consisting of two junctions of dissimilar metals.
(b) Measurement terminals of a voltmeter as an example of (a).

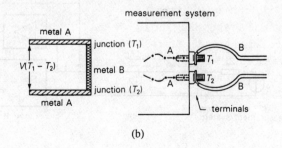

(a)   (b)

are especially easily troubled by errors originating from *leakage currents*. Usually leakage currents affect only the sensitivity of a measurement system. Leakage currents cause *multiplicative errors* as they give rise to an extra equivalent loading impedance.

The order of magnitude of leakage is illustrated by the fact that some time after manufacture, the resistance between two points 1 cm apart on a printed circuit board is usually not larger than 10^8 Ω.

To keep this leakage small, one can take certain measures, for instance, treatment with a dust-proof encapsulation. To minimise the effects of humidity, a circuit can be treated with a water-repellent agent (silicon compound) or the circuit can be placed in a waterproof housing with a hygroscopic substance (silica gel).

A frequently used remedy for leakage currents is to redesign the measurement system input in such a way that the voltage appearing across the leakage resistance is reduced to zero. One method of achieving this is *active guarding* (see Fig. 2.53). The active guard not only eliminates *resistive leakage*, but also '*capacitive leakage*'. In fact, all leakage impedances which are 'cut' by the guard shield are eliminated when the potential of the shield follows that of the guarded conductor. In Fig. 2.53 the leakage impedance Z_l is cut by the guard shield surrounding the non-earthed conductor. If the potential of the guard shield is driven from a voltage follower with a transfer $1 - \varepsilon$ (ε is the relative deviation from the ideal unity transfer) only a fraction ε of the original voltage will appear across Z_l. The impedance Z_l has seemingly increased to a value Z_l' where:

$$Z_l' = \frac{Z_l}{\varepsilon}$$

If, for example, the active guard followed with an inaccuracy of 1%, the leakage impedance would increase by a factor of 100.

Figure 2.53. Active guarding used in a connecting cable to suppress the leakage current through the finite impedance Z_l between the two conductors of the cable.

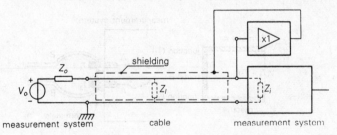

Obviously, ε may never become negative. This would produce a positive feedback loop in the system and would make the system unstable.

On printed circuit boards, often an auxiliary trace is used on either side of the trace which is to be guarded. The two copper traces are then connected at one single point to the output of the voltage follower. To obtain a large increase in the value of the leakage impedance, the guard shield must extend as far as possible into the measurement object and into the input of the measurement system.

Capacitive injection of interference

A certain non-zero parasitic capacitance C_p will always exist between the input of a measurement system and any nearby wires carrying an AC voltage. This will inject an interference voltage into the input circuit of the measurement system (see Fig. 2.54). A common example of this kind of interference is the 50 or 60 Hz power line hum. The interference voltage at the input of the measurement system increases as the 'impedance level' of the input increases; so as $Z_s // Z_i$ increases. Capacitive injection of interference is a form of additive disturbance injected from the environment.

There are several methods of reducing this capacitive injection of interference. One can enlarge the distance to the source of the interference (and thus reduce C_p). One can also lower the impedances of the input circuit. This can be seen as follows. The interference voltage is V_d. If C_p offers a much larger impedance than the input circuit $(Z_o // Z_i)$ we may calculate the interference voltage across the measurement system input as:

$$V_i = v_d \, j\omega C_p \, (Z_o // Z_i)$$

Clearly if $Z_o // Z_i$ is reduced, so is V_i. One can also shield the input circuit by means of an earthed conducting shield (made of copper or aluminium as in Fig. 2.55).

For most purposes a single layer copper shield will provide sufficient protection from capacitively injected interference; the copper screen

Figure 2.54. Capacitive injection of interference into the input of a measurement system.

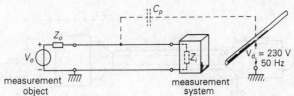

intercepts electrical fields very well. Again, it is important that the shielding is complete. In other words, the shield of the measurement object and measurement system must extend as far as possible, and there must be no gaps or slots. For high-frequency interference especially, one must ensure that the shielding has no open seams, etc.

Strictly speaking, a shield may only be used for shielding and not for the returning earth or ground connection. Fig. 2.55 shows the proper use of a so-called twin-ax cable. The shield conductor must be connected to earth at only one open point. Otherwise large stray currents may flow in the shield due to a difference in the earth potential between two far-away points. These stray currents will inject an interference voltage into the input circuitry. The shield of the cable must be earthed (grounded) at the end of the cable which is attached to the circuit with the lowest impedance. For a voltage measurement system this is always the measurement object side (since $Z_o \ll Z_i$), for a current measurement system this is always the measurement system side (since $Z_o \gg Z_i$). In this way, the potential difference between the shield and the two conductors is kept as small as possible, which minimises capacitive injection from the shield into the internal cable conductors.

Inductive injection of interference
If the (input circuit of a) measurement system is located in an environment with a varying magnetic field an interference voltage will be induced in the input circuit. The varying magnetic field is usually generated by an AC current flowing through a conductor, or by moving magnetic machine parts. Of course, the same effect occurs when the input leads of the measurement system are moving (vibrating) in a constant magnetic field. Fig. 2.56 schematically illustrates the inductive injection of interferences. The induced interference is a form of additive disturbance. The induced disturbance voltage V_d is equal to:

Figure 2.55. Shielding effectively against capacitive injection of interference.

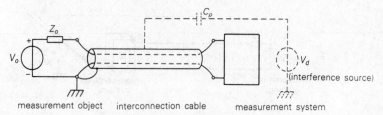

measurement object   interconnection cable   measurement system

$$V_d = -\frac{d\phi}{dt} = -\frac{d}{dt}\int_A B\,dA$$

Assuming the induction B is uniform across the area of the input wire loop, we find:

$$V_d = -\mu_0 A\frac{dH}{dt}$$

If we can express V_d in complex numbers (sinusoidal interference), the interference voltage across the input terminals of the measurement system is given by:

$$V_d' = V_d\frac{Z_i}{Z_i + Z_o}$$

The voltage V_o' which is generated by the measurement object amounts to:

$$V_o' = V_s\frac{Z_i}{Z_i + Z_o}$$

Consequently, the ratio of the measurement signal to the interference V_o'/V_d' cannot be improved by changing the impedances Z_i and Z_o. This obviously holds true only when V_o' and V_d' have the same frequency.

The above expression shows that the induced interference voltage increases as the flux threading the area A increases, as the magnetic field strength $H(t)$ increases, as the field strength changes more rapidly in time, or as the primary current $I(t)$ (Fig. 2.56) becomes larger or changes faster in time.

The inductively induced interference voltage can be minimised in several ways. The magnetic field strength near the measurement system can be reduced, for instance, by placing the measurement system further away

Figure 2.56. Inductive injection of interference in the input of a measurement system.

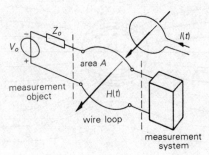

from the source. The spatial orientation of the measurement system can also be changed, such that the enclosed flux is minimal. In Fig. 2.56 this has been achieved by placing the two wire loops in planes perpendicular to each other. In addition, the area A can be minimised by twisting the two conductors (twisted pair). Also, the system can be *shielded* using a ferro-magnetic shield with a high relative permeability μ_r.

The effectiveness of magnetic shielding is best expressed by the shielding factor F. If the (magnetic) field strength is H_0 without shielding and H_s with shielding, the shielding factor is given by:

$$F = \left| \frac{H_0}{H_s} \right|$$

The shielding factor depends on the geometry of the shield as well as on the material used. Fig. 2.57 indicates how the magnetic field is modified when a cylindrical shield is placed in a uniform magnetic field.

The magnetisation of a magnetic conductor increases non-linearly with the applied external magnetic field, and, therefore, the shielding factor depends strongly on the field strength. In Fig. 2.58 the shielding factor F is plotted for three different ferromagnetic materials as a function of the field strength H_0 under the same conditions (same shape and uniform field). The sudden drop of the $F(H_0)$ curves is caused by the magnetic saturation of the shield.

To ensure good shielding, even in very strong magnetic fields, the shield must be thick. For thick shields, however, it is possible to save on the amount of ferromagnetic material used by instead using several, thinner

Figure 2.57. Magnetic field in the presence of a cylindrical magnetic shield placed in a transverse, uniform magnetic field H_0.

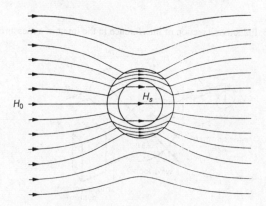

shields, separated by an air gap. In this way, the same shielding factor F can be achieved with less material.

Since a ferromagnetic shield usually has not only a high magnetic conductivity, but also a high electrical conductivity, eddy currents will be generated within the shield when it is placed in an alternating magnetic field. Because of these eddy currents, the external field cannot penetrate as easily into the deeper layers of the shield; the eddy currents act as it were as a shield for the deeper layers. Only the outer layer of the shield participates in conducting the magnetic field. The magnitude of the eddy currents increases with increasing frequency, so magnetic conductivity is repelled to the outer layer of the shield as the frequency of the magnetic field increases. As a result of this so-called magnetic skin effect, the effective depth of penetration of an external field (skin depth δ) decreases as the frequency increases:

$$\delta = \sqrt{\frac{\rho}{\pi f \mu_r \mu_0}}$$

In this expression ρ is the electrical resistivity of the material. The skin depth δ is the thickness of the equivalent shield required to exhibit the same magnetic conductivity as the actual shield with the same μ_r but $\rho = \infty$: in other words, an equivalent shield in which no eddy currents occur. As the skin depth decreases, the magnetic resistance and also the shielding factor F increase, as is shown in Fig. 2.59.

Figure 2.58. Shielding factor F as a function of the field strength H_0 for different ferromagnetic materials with a 1 mm thickness, at 50 Hz. M 1040: $\mu_{ri} \approx 40\,000$, mu-metal: $\mu_{ri} \approx 25\,000$, permenorm: $\mu_{ri} \approx 5\,000$. Here μ_{ri} is the relative initial permeability.

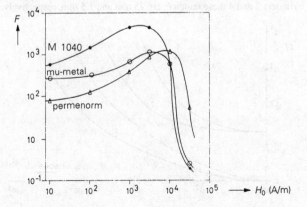

To create a ferromagnetic shield for an alternating magnetic field with a small skin depth δ and a large shielding factor F, a material with a high electrical conductivity (small ρ) and a high magnetic conductivity (high μ_r) is required. For this reason, shields with a high electrical conductivity and with a high magnetic conductivity are often used paired together for multiple layer shielding.

The skin depth δ decreases and the shielding factor F increases with increasing frequency. Therefore, for most high-frequency applications ($f > 10^5$ Hz) a sufficient shielding factor can be obtained with only an electrically conductive shield.

Injection of interference by imperfect grounding
In a measurement set-up it may occur that the measurement object is grounded at a different point on the earth rail than the measurement system: for instance, when two different outlets of the mains supply are used. Such a situation is shown in Fig. 2.60.

Equipment grounded in this manner can pick up considerable interference from stray ground currents. Usually the resistance R_g of the ground rail between the measurement object and the measurement system (of the order of 0.1 Ω/m) cannot be neglected, This non-zero resistance and the stray ground currents in the ground rail (caused by other equipment grounded at the same rail) will then produce a voltage across R_g, in series with the source voltage V_o. This additive disturbance will be relatively large when small signals have to measured and can easily throw off sensitive measurements.

Figure 2.59. Shielding factor F of a cylindrical shield placed in a transverse magnetic field, as a function of the frequency. Curves 1 and 2: steel cylinder, $\mu_r = 300$, $\rho = 0.17\ \mu\Omega$ m; curves 2, 3 and 4: copper cylinder, $\mu_r = 1$, $\rho = 2\ \mu\Omega$ m. Cylinders 1 and 3 have a diameter of 50 mm and a wall thickness of 3 mm. For cylinders 2 and 4 these numbers are 25 mm and 1.5 mm, respectively.

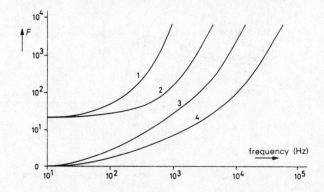

A solution which is frequently used to avoid this stray current interference is the strict adherence to *single point grounding* for both measurement object and measurement system (depicted in Fig. 2.61). The voltage across R_g is here kept 'outside' the input circuit of the measurement system and has therefore no effect. For most measurements, this is the preferred method of grounding and provides adequate elimination of stray current interference. However, a small interference voltage will still be present, as this system is now grounded via the conductor AB. This conductor carries the ground current of the measurement system. The prevalent cause of this instrument ground current is the capacitive coupling between the primary and secondary windings in the transformer of the power supply of the instrument, which can give rise to a considerable 50 or 60 Hz AC current. However, this instrument ground current is usually much smaller than the ground return current in Fig. 2.60. Even so, since the resistance R_{AB} is in series with the signal source, this interference can still result in considerable hum at the input terminals of the measurement system.

To further avoid interference originating from imperfect grounding it is advisable to use a measurement system with a *symmetrical input* with respect to earth. Such an input is also referred to as a *floating input* or *differential input*. We will return to this type of measurement system in

Figure 2.60. Grounding the measurement object and the measuring system at different points on a ground rail.

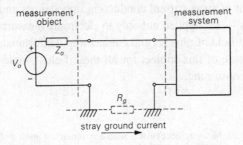

Figure 2.61. Single point grounding of the input of a measurement system.

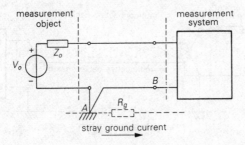

Section 3.3.4. It is commonly used when measuring very small input signals. These differential measurement systems measure only the potential difference between the input terminals (denoted by + and – or 'Hi' and 'Low' in Fig. 2.62). They are grounded via a separate terminal (terminal 0 in Fig. 2.62). This cabinet or case grounding is necessary for safety reasons. The ground current of the measurement system will flow through this separate terminal and will therefore remain outside the input circuitry. The voltage across the ground rail resistance R_g is superimposed on the potential at both the + and the – input terminals (with respect to the 0 terminal). Such a differential measurement system is designed specifically to be highly insensitive to any voltage common on the two input terminals.

The insensitivity of the measurement system is referred to as 'rejection'; the insensitivity of the system to potentials common to both the plus and minus terminals is called the 'common mode rejection'. The system sensitivity to the potential difference between the + and – terminal is the 'differential-mode sensitivity'. Also see Section 3.3.4.

Fig. 2.62 also shows how to avoid capacitive injection of interference by shielding the cable with a conductive sheath, grounded at the side of the measurement object (voltage measurement). It further shows how to avoid inductive injection of interference by using a cable with twisted inner conductors.

The methods for eliminating interference which have been considered above — i.e. the application of active or grounded shielding, exclusion from the input circuit of stray current conducting impedances and compensation of disturbances — are applied not only to electronic measurements, but also to almost every field of physics and engineering. Unfortunately, however, a general discussion of this subject for all these fields would easily become long and time consuming.

Figure 2.62. Measurement system with a symmetrical input with respect to ground and a shielded connecting cable with twisted internal conductors.

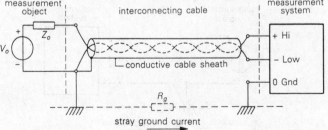

2.3.3.4 Observer influence: matching

In Section 2.3.3 we have seen that the interaction between the measurement system and the observer comprises two components. The desired component is the influence of the measurement system on the observer, which transfers the measurement information to the observer. The undesired component is the influence of the observer on the measurement system. Unfortunately, the latter component often causes the measurement result to depend on the observer.

The measurement result of a dial instrument, for instance, is presented as the deflection of a pointer in front of a fixed calibrated dial. If there is some distance between the dial and the pointer, the true indication is the value at the scale on the dial at the orthogonal projection point of the pointer onto the dial. If an observer does not read the pointer position with one eye held in a plane through the pointer perpendicular to the dial, then the observer, in fact, alters the measurement system (oblique projection instead of a orthogonal projection). Therefore, the measurement result is also different (parallax error). The observer has introduced a reading error.

In addition to incorrect observations (*reading errors*), errors can also occur due to the observer influencing the operation of a measurement system (*adjustment errors*). In other words, influencing errors occur at the 'interface' between the observer/operator on the one hand and the measurement system on the other hand.

Example: When performing an impedance measurement with a continuously adjustable bridge circuit, the observer/operator will make both reading errors (reading the null detector and the dial of the controls for adjusting the bridge impedances) and adjustment errors (zero adjustment of the null detector, balancing the bridge with the bridge variables). The influencing error of the observer/operator in this example will mainly be a reading error, if the sensitivity of the null detector to a variation of the bridge variables is held small. Then, however, in order to make only a small error, the null detector must be read extremely precisely. If the sensitivity of the null detector is made large (a small imbalance of the bridge variable causes a large deflection), the bridge must be adjusted extremely accurately. The influencing error is now mainly an adjustment error. (For the sake of simplicity, the reading error which is made afterwards, when reading the dial of the bridge impedance, is not considered in this example.) In practice, one tries to keep reading errors and adjustment errors down to approximately the same small value (optimal measurement).

To keep the influencing errors of the observer/operator small, we have to adapt the measurement system to the skills of the observer. This is done by adapting the parts of the measurement system which actually interact with the observer (displays) or operator (operating panel, knobs, etc.) to the specific characteristics expected from observers/operators.

In Section 2.3.3 we generalised the term observer/operator, to included both human and (machine) automatic observers and operators (for instance, computers). When submitting a measurement result to a machine, it is possible to match the output section of the measurement system, or, *vice versa*, to match the input section of the machine to the output of the measurement system, to minimise influencing errors (such as load mismatch errors). In the same way, it is also possible to match the output section of a machine to the input section of a measurement system, to control (operate) the measurement system better. Therefore, matching at the interface of an automatic observer/operator and a measurement system almost never gives any problems.

As far as the human observer/operator is concerned we can remark that, after a certain period of training, fewer errors will be made. Humans learn how to adapt to a specific measurement system. The purpose of adapting a measurement system to a human observer/operator is to enable unskilled observers to obtain correct measurement results immediately and to make the observations and adjustments less tiring, so that the chance of making mistakes and errors is reduced.

In the following discussion we will confine ourselves to matching the measurement system to a human observer/operator. Here, the measurement system must present the measurement result in such a way that the impression gathered by the senses of the observer is identical to the information contained in the measurement result. The study of this presentation is part of *ergonomics* and *sensory physiology*. Further, the controls of a measurement system must be designed to prevent the occurrence of adjustment errors as far as possible.

Almost all measurement instruments rely exclusively on sight as the only one of our five senses (sight, hearing, touch, smell and taste) for transferring information. Therefore, an electrical signal at the output of an electrical measurement system must be converted into an optical signal. This is accomplished by an output transducer: the *display*. The human operation and adjustment of an instrument is almost always performed manually, and consequently, the controls of the instrument (knobs, buttons, etc.) must be well adapted for operation by hand, i.e. our tactile skills, our strength and the size of our fingers.

A few examples of the considerations necessary for an instrument to be operated conveniently are summed up here: The controls must be positioned in functional and logical groups in front of the right shoulder of the (right-handed) operator. The controls and switches must not be too small or placed too close together. For a continuous adjustment which requires little force, a finely ribbed knob must be used. If the setting of a control is important, the control should have an arrow shaped knob or a window dial.

Considerations for convenient visual observation are: The shape and size of the symbols used on the display must guarantee the legibility of a measurement result; the background contrast must be large enough, with preferably black letters on a white background, or *vice versa*. The type of dial must be suited for the purpose of the display. For instance, if two values are to be compared, neighbouring vertical gauges dials should be used. For reading measurement values horizontal gauges are preferred. For reading rapidly fluctuating values circular dials are indicated. The deflection of the pointer must correspond to an increase of the measured quantity as follows: a horizontal gauge; from left to right: a vertical gauge; from bottom to top: a circular gauge; clockwise.

We distinguish two types of display; analogue or continuous displays and digital or discrete displays. An *analogue display* can indicate any value between given upper and lower limits; the indication range is continuous. This form of display is called analogue since the indicator moves analogically to the input quantity. The range of a *digital display* is discrete; between given upper and lower limits only a certain finite number of values can be displayed.

Most analogue displays produce a similar reading: usually a rotational or translational movement with respect to a fixed reference grid or scale (see Fig. 2.63). The magnitude of the deflection with respect to the background reference is a measure of the magnitude of measurand. For instance, the deflection of an electron beam with respect to the grid on the face plate of a cathode-ray tube, or the movement of the horizon symbol of an artificial horizon in an aeroplane.

The most commonly used digital displays for measurement purposes simply produce a series of symbols (numbers, arithmetic symbols, letters, etc.). Examples of these digital displays are numerical displays which only display numbers (see Fig. 2.64(a)) and alpha-numerical displays which, in addition, can also display other symbols (see Fig. 2.64(b)). Both display examples in Fig. 2.64 are used in electro-optical displays.

Besides analogue and digital displays there are also hybrid displays. These are, in fact, digital displays in which the least significant decimal is displayed in a continuous fashion, for instance, the odometer of a motor vehicle (see Fig. 2.65).

An analogue display not only indicates well the *magnitude* of a deflection, but also provides a good impression of the *trend* in the time of the deflection. The trend dy/dt of a deflection $y(t)$ is determined by the velocity and direction of change in $y(t)$. A digital display, on the other hand, is not suitable for showing the trend in the measurand. Thus, an analogue display ·

Figure 2.63. Traditional analogue displays. (a) Fixed gauges with a rotating or translating pointer. (b) Complex analogue display (artificial horizon).

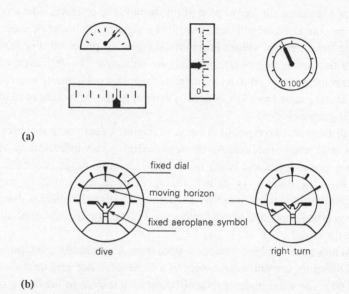

(a)

fixed dial

moving horizon

fixed aeroplane symbol

dive

right turn

(b)

Figure 2.64. Segment displays with (a) seven segments and (b) a matrix display consisting of 5 × 7 dots.

(a)

(b)

Figure 2.65. Hybrid display. The first four digits jump from one value to the next, but the last digit moves continuously.

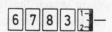

constitutes not only a *magnitude display*, but also a *trend display*. This has several advantages, especially when setting, aligning or adjusting an instrument. This can usually be done more quickly when using analogue displays. (A null detector must therefore have an analogue read-out.) However, an analogue display does unavoidably imply the introduction of *reading errors*. Digital displays do not introduce such errors. This means that even if the indication along the scale of an analogue meter is exactly correct, the precise value can still not be read-out.

When an analogue meter is read-out, the observer will assign a certain scale value to the deflection of the pointer. The observer is then, in fact, acting as an analogue-to-digital converter. However, in the process the observer will make errors (parallax errors, interpolation errors and so on), which may be assumed to be distributed randomly around the true value. Unfortunately, the reproducibility of the human analogue-to-digital converter is rather poor. Therefore, the obtainable accuracy with an analogue display is low (approximately only 1–0.1% of the full-scale value).

When reading the measured value of an analogue display, errors are introduced (parallax errors, interpolation errors and — if the instrument needs to be nulled before the measurement — offset errors). To minimise these errors, it is important to match the instrument to the observer. Large and clear dials with a linear scale, and a subdivision of the scale in 10 or 3 main subsubdivisions make interpolation easier. A reflective dial will considerably reduce parallax errors.

Digital displays present a number to the observer, so no reading errors can occur. (Of course the observer can still make *mistakes*.) Therefore, digital displays are extremely well suited for displaying constant quantities with a high accuracy. The only error introduced by the use of digital displays is a *quantisation error*, due to the finite resolution of the display. Most digital displays simply drop the least significant decimal, causing a truncation error. The magnitude of this error is the reciprocal value of the indicated number (ignoring the decimal point).

As we have seen the reading of an analogue display is inherently afflicted with reading errors, whereas that of a digital display will always be associated with quantisation errors. The measurement system will produce its own errors (errors arising in the system, and also disturbances from the environment). These errors are often referred to as *display errors*. It is obvious that it is pointless to design a system with a display error which is smaller than the reading error or quantisation error.

The reading errors made with an analogue display which has a linear scale are constant errors: they have a constant value over the entire range of the instrument. The absolute reading error is therefore independent of the magnitude of the deflection. This error is therefore given as a percentage of the *full-scale value*. Display errors originating from, for instance, a deviation of the sensitivity of the measurement system from the nominal value will cause a relative error, whose magnitude is constant across the entire linear scale of the display. Therefore, this error is stated as a percentage of the *indicated value*. The example in Fig. 2.66 (with, let us say, a display error of 0.5% of the indicated value, and a reading error of 0.5% of the full-scale value) shows that over the range from 30% to 100% of full scale too wide an error specification would be obtained if the accuracy were specified as a percentage of the full scale alone, or as a percentage of the indicated value alone. The display would be, in fact, far more accurate.

2.4 Structure of measurement systems

Until now, we have always referred to the complete collection of measurement devices and instruments needed for a measurement as the measurement system. 'System' can here refer either to a single apparatus or to an elaborate measurement set-up. We did not detail the internal configuration or structure of the measurement system. In this section we will discuss the functional internal structure of a measurement system (see Fig. 2.67).

The information we wish to acquire from the measurement object is not always available in the form of active information. Where the measurand is

Figure 2.66. Relative error $\Delta y/y$ of the deflection y for three different error specifications of an analogue display with a linear scale.

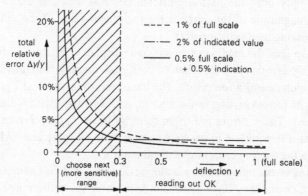

not active, we must use an *exciter*, which supplies a stimulus to the measurement object. The response of the object (together with the stimulus) now contains the desired information. If, however, the measurement object emits a signal which already contains the desired information, external excitation is not necessary.

Often, the parameter or variable we wish to measure is an electrical one. If one has to measure non-electrical parameters or variables — such as stiffness, thermal resistance, displacement, etc. — some kind of converter or *transducer* rather than a completely mechanical or thermal measurement system is often used. The transducer will convert the input parameter or variable into an electrical output signal, which carries the information of the original measurand.

A great advantage of this conversion into electrical signals is the fact that this enables us to process the information further electronically; this is a very inexpensive and flexible means of processing. The information can, for instance, now easily travel large distances, with minimal environmental disturbance. By *transmitting* the measurement information we can measure remotely (referred to as *telemetry*). This is especially useful for measurements in inaccessible or hostile places, such as underwater (oceanography), in the atmosphere (meteorology), or for measuring a large number of objects which are spread far apart (e.g. measurement in food and oil processing industries).

Sometimes, measurement information is transmitted by other than electrical means. In some processing industries which deal with flammable substances pneumatic telemetry is used for transferring information. Measurement data are here transmitted through a thin pipe in the form of the pressure of a gas.

Figure 2.67. Generalised internal structure of a measurement system.

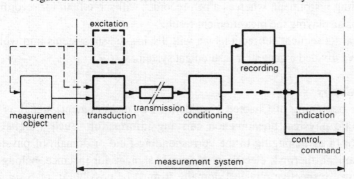

If a telemetry channel is interrupted (short-circuited) it may, in some cases, cause dangerous situations as suddenly a zero value will appear at the output although the measured value is not zero at all. For this reason, a particular quiescent current or voltage is often defined as the zero value, a so-called *live zero*. A failure is detected when this live zero is interrupted, and a fail safe mechanism can be activated to revert to a safe condition.

In a telemetric system, measurement information is not necessarily represented by the amplitude of an electrical potential or the magnitude of a current. To make the system more insensitive to disturbances, the information is sometimes contained in the *frequency* of a signal or in the *width* of the *pulses* of signal with constant frequency, or it may even be transmitted digitally. Such transmission methods give a better noise and interference immunity.

The electrical output signal of a transducer is usually not suitable for direct presentation to an observer. Often, it must first undergo several stages of *signal conditioning* (such as amplification, filtering, correction for non-linearity of the transducer, etc.).

After this conditioning the signal can be presented to an observer. We display the result to a human observer, we *control* a mechanical (automatic) observer with the resulting output signal. The output signal may also be stored temporarily in memory and used later. Then, we speak of *recording* the measurement result.

Not all of the six subsystems in Fig. 2.67 are present in every measurement system. The subsystems need not necessarily occur in the illustrated order. Often, for instance, there will be some form of signal conditioning before transmission of the signal.

The various subsystems may sometimes be incorporated in a single measurement instrument, but they can also be implemented separately, as individual instruments. A magnetic tape recorder, for example, is only a recording instrument, whereas a pen recorder will serve both for recording and for displaying the measurement result.

In the subsequent subsections we will discuss the subsystems into which we have divided a general measurement system.

Transducers

At the beginning of Chapter 2 we defined a (measurement) signal as an energetic physical phenomenon carrying information. Such a signal is thought of as belonging to the corresponding field or domain of physics. Mechanical, thermal, electric or magnetic signals, for instance, belong to their own respective physical domain. It must be possible to map signals from one physical domain onto signals of another domain, in order to implement the transfer of information from one domain into another (see Chapter 1). Such mapping is realised by 'converters' which can transfer one kind of energetic physical phenomenon (in one domain) into another kind (in another domain). The transfer must preserve the information contained in original energetic phenomenon. These information preserving energy converters are called *measurement transducers*.

In addition to being able to map signals from different domains onto each other, it can also be necessary to map signals from the same domain onto each other. In this case, an energetic phenomenon is converted into a similar energetic phenomenon while preserving the relevant information of the original phenomenon. This may be necessary to increase the power of a phenomenon (power amplification), or to omit certain irrelevant information (filtering). We will discuss these converters extensively in Section 3.3, which deals with electronic signal processing.

Physical effects in materials which are used for mapping signals of different domains are referred to as cross effects, whereas the effects in materials which are used for mapping signals within the same domain are called direct effects. Examples of cross effects are: from electrical to thermal — the Peltier effect; from thermal to electrical — the Seebeck effect; and from magnetic to electric — the Hall effect. Examples of direct effects in materials are: in the electrical domain — electrical resistance; and in the mechanical domain — elasticity.

In relation to the energy conversion property of transducers, we distinguish two types of transducer: passive and active.

Passive transducers are transducers which operate without the supply of auxiliary power (see Fig. 2.68(a)). The (average) power P_o at the output is drawn from the measurement object in the form of the (average) power P_i. The physically realisable energy conversion will always involve losses (the power P_l), however, and therefore:

Figure 2.68. Passive and active transducers. P_i is the input power, P_o is the output power, P_l is the power which is lost in the conversion process, P_c is the control power and, finally, P_{ps} is the auxiliary power supplied to the transducer. (a) Passive transducer. (b) Active transducer.

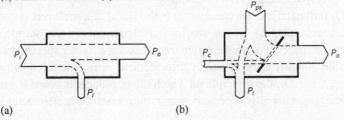

(a)   (b)

$$P_i = P_o + P_l$$

It is possible that during a given short time interval, energy is stored within the transducer. Therefore, the above expression only holds for the average values over a longer period in time. If the measurement object may not be significantly loaded and, consequently, only a very small input power is available, the *efficiency* η of the passive transducer becomes relevant:

$$\eta = \frac{P_o}{P_i} = \frac{P_i - P_l}{P_i} = 1 - \frac{P_l}{P_i}$$

The efficiency of the conversion process is, of course, less important when more input power is available. All mechanical volt meters, ammeters and watt meters are examples of this class of passive transducer. They convert electrical energy into mechanical energy in the form of the potential energy of the armature of the meter against a compressed spring.

Active transducers are transducers which require an auxiliary supply power (see Fig. 2.68(b)). The output power P_o of the transducer is supplied almost entirely by this auxiliary supply power P_{ps}. The power which the measurement object delivers is almost zero. Only a small amount of power P_c is required for controlling the transducers output power (in Fig. 2.68(b) this conversion process is schematically indicated by a gate valve). By supplying auxiliary power, amplifying transducers can be realised with a very high sensitivity.

As we have stated many measurement systems employ transducers for converting non-electrical signals into electrical signals, because conditioning, processing and transmission of signals in the electrical domain are relatively straightforward. We will therefore limit our discussion to transducers which map signals of different physical domains onto the electrical domain and to reverse transducers which convert an electrical signal into a non-electrical quantity. The first category of transducers is required at the *input* of a measurement system. They are therefore called *input transducers* or *measurement transducers*. The second category is required at the output of the measurement system for displaying or recording data or for controlling other processes. They are therefore referred to as *output transducers* or *actuators*.

Unfortunately, the nomenclature in the world of transducers is not unique. Various names by which one indicates transducers are: sensor, sensing element, pick-up or gauge. Often transducers are further specified according to the quantity they measure: displacement sensor, accelerometer, strain gauge, etc. Or, the principle on which the transducer is based is used in the name: capacitive displacement sensor, piezoelectric accelerometer, resistive strain gauge, etc. In Section 3.2 a number of simple, but widely used transducers will be considered.

Signal processing

One of the reasons for converting non-electrical signals into electrical signals is the vast variety and flexibility for processing which is offered by modern electronics. Usually, an electronic measurement signal or the output of a transducer is not immediately suitable for direct display, recording or for controlling a machine and must therefore be conditioned first. The signal conditioning may be linear, frequency dependent (filtering) or (quasi-) frequency independent (amplification, attenuation). Non-linear operations are also possible (rectification, RMS-value detection, analogue-to-digital conversion, etc.). Signal conditioning can be performed in many ways; in Section 3.3 we will study several of the more commonly used methods.

Displays

We have referred to the device that takes care of the presentation of measurement results to a human observer as the 'display'. Displays need not necessarily be *analogue* (for example, the trace an electron beam writes on an oscilloscope screen), but can also be *digital* (such as alpha-numerical displays or signalling lights). We have also seen that the display is designed for visual observation and is, therefore, an electro-optical transducer. Analogue displays especially must be well matched to the observer, to avoid large reading or interpretation errors. This problem is partially solved by using a *hybrid display*. An example of this is an oscilloscope which can write information such as sensitivity, time base, etc. on the screen with the same electron beam that writes the trace. Matching a display to the observer is important for digital displays, to avoid fatigue and any resulting mistakes (minimise reflections and high contrast, use pleasant colours, clearly legible symbols, etc.). Several common displays used in measurement and instrumentation are reviewed in Section 3.4.

Recording data

Data are recorded so that they are available at a later point in time, for instance, for displaying at a later more convenient date. Recording is often used when one collects large quantities of measurement results to facilitate a later evaluation of the results. Take, for example, the 'black box' in an aircraft, which is only read later, when the measurements are relevant (after the occurrence of an accident). Another reason for recording measurement data is to prevent one having to repeat the measurement (for instance, with extremely expensive tests (collision tests)). Graphical recording is often used to assist the *interpretation* of the measurement results. Examples of this are *x–t* and *x–y* diagrams, or polar plots. When measurement results are recorded in this manner, the structure and relation of the measurement results are revealed, which simplifies the interpretation. The electrical activity of the heart, for instance, is recorded in an *x–t* plot: the electro-cardiogram. A specialist can then easily recognise specific patterns or irregularities therein. Data can be recorded in an analogue or in a digital fashion; e.g. graphical recording by means of a pen plotter or magnetic recording on a tape. We will discuss the means for recording measurement data further in Section 3.5.

Control, feedback

Sometimes a measurement result is not used for displaying or recording, but for direct control of a process. The purpose of process control is to regulate the output according to specific requirements. One or more variables of the process are measured and the process is controlled to diminish any difference between the measured values and the preset values. If the control of a process is based on the measurement of a variable associated with the process and the resulting change in the process quantities does not affect this measured variable, we regard the process to be regulated by *feed-forward control* (open-loop control). However, if the control is based on measurements affected by the previous control actions, a closed-loop arises (which, in some cases, potentially can cause instability). This latter method of process control is called *feedback control*.

3

Measurement devices in electrical engineering

3.1 Introduction

Following the more general discussion of the previous chapter on the measurement of physical quantities, we will now zoom in on measurement devices (systems and subsystems as well as components) that are used frequently in electrical engineering. The nature of these devices depends on the objective we wish to achieve with the measurement.

A number of possible objectives are listed below:

- *Research.* The goal of conducting research is to enlarge our insight into all sorts of natural and artificial physical states and phenomena. Therefore, in a research environment, measurement systems typically must be capable of measuring over a large range, with an excellent linearity and good dynamic behaviour.
- *Measurement of a consumed quantity.* Consumption measurements are performed primarily to determine and record the size of a delivered or absorbed quantity. Take, for instance, the electric energy meter (kWh meter) present in every house or an electronic scale in a shop. For this kind of application the most important aspect is the accuracy of the measurement, as the consumer must be charged correctly for the quantity used. Therefore, frequent calibration of the equipment is often required by law.
- *Safety.* Measurements are often undertaken to ensure the safety of humans and their environment, for instance, measurements of the level of nuclear radiation or the concentration of toxic substances in, let us say, drinking water. The reliability of these safety systems is crucial; the system must

always function according to the specifications. If, however, a failure did occur, it must not lead to dangerous situations; the system must be fail-safe.

– *Verification.* Here the objective of the measurement is to determine whether a product will meet certain specifications. This requires regular calibration of the limits of the tolerance interval of these so-called go–no-go measurements.

– *Process control.* In industry, for instance, the objective of many measurements is to obtain information on the state of a given process. Corrections can then be made, based on these measurements. If, as a consequence, the measured process quantities vary, the measurement system is part of a feedback loop. The dynamic behaviour of the measurement system becomes critical here, as this will contribute to the stability or instability of the monitored system.

Usually, a measurement system will convert the measured physical quantity into an electrical signal as soon as possible. This is because an electrical signal can easily be processed into almost any desired form. The variety of available electronic manipulations enables us to realise the necessary signal processing in a fast and inexpensive way.

The rapid development of electronic signal processing has several causes. First of all, electronics circuits can very easily realise *signal amplification*. In the process of amplification the power level of a signal is increased without appreciable loss of information. Therefore, with electronic instrumentation one can easily obtain a high sensitivity. A photo-multiplier device, for example, can easily achieve an electronic current multiplication factor of 10^6–10^8.

Secondly, the use of electronics makes it possible to perform measurements which only minimally load the measurement object. For instance, the power extracted from a fluid when the pH is measured with an electrometer amplifier is less than 10^{-15} W.

Furthermore, electronic circuits have no moving parts, are therefore silent, free of wear and have a relatively low intrinsic energy consumption.

Probably one of the greatest advantages, though, is the *speed* at which electronic circuits can process (quickly changing) phenomena, due to the lack of moving parts with the associated inertia. Even events happening within 100 ps can still be detected. The frequency range of electronic circuits can reach well beyond 10 GHz.

In addition, electronic signal processing offers a great deal of *flexibility*; numerous functions can be realised and intertwined almost limitlessly to create more complex functions. Measurement information can easily be transferred across long distances (telemetry), with a wide bandwidth and very low sensitivity to disturbances.

Electronic instruments do, however, have their *disadvantages*, amongst which are the facts that they cannot process high power signals (for which one needs hydraulic signals), the reliability frequently leaves something to be desired and, finally, the equipment is quite susceptible to such environmental influences such as temperature, humidity, nuclear radiation and so on.

In the following discussion of measurement transducers we will limit ourselves to transducers which convert a non-electrical, physical quantity into an electrical quantity (input transducers) and *vice versa*, to transducers which convert an electrical quantity to a non-electrical, physical quantity (output transducers). We will start with input transducers and deal with output transducers in Section 3.4.

3.2 Input transducers

Before we discuss the transduction principles frequently used for measuring common physical quantities such as displacement, velocity, temperature, magnetic induction, etc., we will first examine several methods for combining individual transduction principles into one single compound transducer. These 'composite methods' are employed to reduce or even totally eliminate certain restrictions associated with individual transducers.

A widespread method of combining transducers is to use two identical transducers in a *balanced configuration* (see Fig. 3.1(a)). If the transducers T and T' both have the same transfer characteristic $y' = f(x')$, the output y of the balanced configuration is:

$$y = f(x) - f(-x)$$

Here $f(x)$ may be a non-linear transfer function that we intend to linearise.

Let's assume that $f(x)$ can be expressed with a Taylor expansion in the following form:

$$f(x) = a_0 + a_1 x + a_2 x^2 + a_3 x^3 + a_4 x^4 + a_5 x^5 \ldots$$

Using the balance equation above we find:

$$y = 2a_1 x + 2a_3 x^3 + 2a_5 x^5 \ldots$$

Evidently, the constant (or offset) a_0 and the even terms a_2x^2, a_4x^4, ... disappear when balancing the two transducers. If the non-linearity of $f(x)$ does not contain any uneven terms, we will obtain a perfectly linear system. The system is then referred to as a '*difference configuration*'. However, usually balancing will only improve the linearity of the system over a limited range of the input quantity x. Such a system is called a '*differential configuration*'. The balanced configuration is not sensitive to external disturbances, since it inherently makes use of *parallel compensation* (see Section 2.3.3.3). The configuration is immune to additive disturbances when the transducers T and T' have the same sensitivity to these disturbances. In order to be immune to multiplicative disturbances, the transducers T and T' must have disturbance coefficients with the same magnitude, but opposite signs. Fig. 3.1(b) shows an example in which a balanced transducer configuration is implemented. The individual transducers are both capacitive displacement sensors, denoted by C and C'. The input quantity is the displacement x and the output quantity is the voltage V_o. Assuming the fringe effects at the edges of the capacitor plates are negligible we find:

$$C = \frac{\varepsilon_0 A}{d + x} \qquad \text{and} \qquad C' = \frac{\varepsilon_0 A}{d - x}$$

Clearly, the relation $C = C(x)$ is not linear. The input quantity x appears in the expression for C' with a negative sign. The output voltage V_o of the transformer bridge can be shown to be proportional to the relative difference in capacitance between C and C':

$$V_o = V \frac{n_2}{n_1} \frac{(C' - C)}{(C' + C)}$$

Substitution gives:

Figure 3.1. (a) Balanced transducer configuration, T and T' are two identical transducers. (b) Application of the balanced configuration in a capacitive displacement sensor.

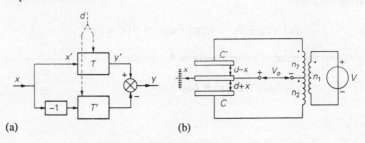

(a)   (b)

$$V_o = V\frac{n_2}{n_1}\frac{x}{d}$$

In this example the application of two capacitive displacement transducers in a balanced configuration achieves perfect linearity. We can therefore write:

$$V_o = x S$$

The sensitivity S is given by $S = Vn_2/n_1d$. The bridge supply voltage V determines the sensitivity.

A second commonly used configuration is the *feedback configuration* of two transducers T_1 and T_2 (see Fig. 3.2). The purpose of the system is to convert an input signal x into an electrical output signal y. We could use a single transducer T_1 for this. Assume, however, that T_1 is unsuitable to be used directly, due to an unacceptably large non-linearity and has too large a sensitivity to disturbances. If we have a second transducer, capable of the reverse conversion (the conversion of y into x) and this conversion is linear and not susceptible to disturbances, then when we combine both transducers T_1 and T_2 (with an amplifier A to increase the loop gain) in the feedback configuration of Fig. 3.2(a), we can realise a compound transducer for conversion of measurement signals from x into y, with the same characteristics as the employed reverse transducer (see Section 2.3.3.3). The necessary conditions for achieving this are a large loop gain and a quasi-static dynamic behaviour for T_1, T_2 and A. In practice, though, the dynamic behaviour of especially T_1 and T_2 is often of a higher order and, therefore, the situation may not be as ideal as described above.

Fig. 3.2(b) shows the application of the feedback configuration in an accelerometer. The input quantity, the acceleration x, exerts a force on the seismic mass attached to the 'voice coil' of an electrodynamical output

Figure 3.2. (a) Negative feedback configuration. T_1 performs the desired conversion of x into y and T_2 performs the reverse conversion. (b) Application of the feedback configuration to an accelerometer.

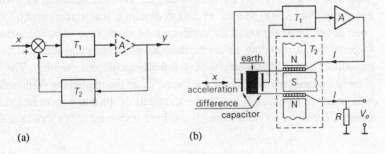

(a) (b)

transducer. The acceleration causes the mass to move. This movement is detected by a displacement sensor (here, a difference capacitor). The output signal of the displacement sensor is amplified and supplied to the coil of the output transducer T_2 in the form of a current, counteracting the motion of the seismic mass. Therefore, when the gain A is large, the mass will hardly move. Thus, the transfer characteristic of this accelerometer is determined by the relation between the current I through the voice coil and the force which the coil exerts on the seismic mass sensing the acceleration. The current I can be converted into an output voltage V_o by a resistor R.

In this type of transducer, the input quantity is automatically compensated by an internal quantity by means of a negative feedback circuit. Therefore, it is often referred to as a force-balance transducer.

The input range of most transducers is small, the dynamic range is often no more than a factor of 3–10. For those cases where this is inadequate, a number of transducers with different input ranges can be connected together to provide a wider input range. If the measured quantity exceeds the input range of one transducer, the following transducer takes over. In this '*relay configuration*' all transducers must have the same sensitivity, otherwise the overall transfer characteristic contains discontinuities. Also, the individual transducers must withstand an input overload very well.

Finally, the reliability of a transducer is often a problem. Sometimes this can be solved by using several transducers which all measure the same quantity. As long as the differences between the outputs of the transducers remain within a certain given tolerance interval, the overall measurement result is given as the average of the individual output signals. However, if one of the outputs differs substantially from the rest, the corresponding transducer is eliminated from the output. If n transducers are used, $n-2$ transducers can malfunction before the system actually becomes defective. This '*redundant configuration*' greatly improves the reliability of the system at, of course, an increased cost.

Before dealing with the various principles of transduction, a final useful remark must be made. The dynamic behaviour of a linear transducer can be described simply by its electrical analogue. We will illustrate this with an example. Fig. 3.3(a) shows an electrodynamic transducer which can be considered linear for small excursions of the membrane. Assuming that the force of the surrounding air on the membrane varies sinusoidally (using the transducer as a microphone), we can introduce complex numbers. The force F on the membrane will cause a velocity V_m of the membrane. This velocity will generate a voltage V across the terminals of the coil which, in turn, will cause a current I to flow through the load impedance. We can regard the transducer as a linear four-terminal network which is inhomogenous; this means that corresponding quantities (across and through) have different dimensions at the input and the output sides of the network. The relation between the input and output quantities is given by the following two transmission equations:

$$V_m = t_{11}V + t_{12}I$$

$$F = t_{21}V + t_{22}I$$

For the time being we will assume that the transducer is ideal. So the coil has no electrical resistance, capacitance or self-inductance and the membrane has an infinite mechanical compliance, no mass and no friction. Applying Faraday's induction law for a coil in a constant induction B results in $V = nlBV_m$, in which n is the number of turns and l is the length of a single turn. The Lorentz force acting on the coil is equal to $F = nlBI$, so:

$$\begin{bmatrix} V_m \\ F \end{bmatrix} = \begin{bmatrix} \dfrac{1}{nlB} & 0 \\ 0 & nlB \end{bmatrix} \begin{bmatrix} V \\ I \end{bmatrix}$$

Since the determinant of this matrix is equal to 1, the transducer must be reciprocal. It can be used as an input transducer (microphone), but also in reverse as an output transducer (loudspeaker). If it is used as an input transducer and the coil is not loaded by a termination impedance, the output voltage will be proportional to the velocity of the membrane. If the trans-

Figure 3.3. (a) Electrodynamic transducer. (b) Linear inhomogenous four-terminal network as an analogue of (a). (c) Three cascaded four-terminal networks. (d) Electric analogue of (a).

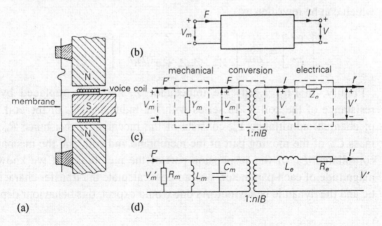

ducer is used as an output transducer driven with a certain current and the membrane is held still, so that the velocity is negligible, the force exerted on the membrane top will be proportional to the current (also see Fig. 3.2(b)).

Apparently we can consider this ideal passive transducer as a transformer which transforms the dimension of the input quantity into that of the output quantity. The transfer ratio of this dimension transformer is $1 : nlB$. Taking into account the impedance Z_e of the coil, we find with Fig. 3.3(c):

$$
\begin{bmatrix} V \\ I \end{bmatrix} = \begin{bmatrix} 1 & Z_e \\ 0 & 1 \end{bmatrix} \begin{bmatrix} V' \\ I' \end{bmatrix}
$$

Even when the coil is not loaded by a termination impedance it will still take a certain force to give the membrane a certain velocity, due to the mechanical impedance Z_m of the moving part of the transducer. For this we can write: $Y_m = 1/Z_m = V'_m/F'$ when $I' = 0$.

This demonstrates one of the disadvantages of across and through quantities. The mechanical impedance is equal to the *reciprocal* value of the ratio of the V and the I quantities (see also Section 5.4). This is due to the arbitrary choice of impedance in the domain of mechanical engineering.

The associated transmission equations therefore become:

$$
\begin{bmatrix} V'_m \\ F' \end{bmatrix} = \begin{bmatrix} 1 & 0 \\ Z_m & 1 \end{bmatrix} \begin{bmatrix} V_m \\ F \end{bmatrix}
$$

Therefore the transmission equations for the entire transducer result in:

$$
\begin{bmatrix} V'_m \\ F' \end{bmatrix} = \begin{bmatrix} 1 & 0 \\ Z_m & 1 \end{bmatrix} \begin{bmatrix} \dfrac{1}{nlB} & 0 \\ 0 & nlB \end{bmatrix} \begin{bmatrix} 1 & Z_e \\ 0 & 1 \end{bmatrix} \begin{bmatrix} V' \\ I' \end{bmatrix}
$$

which can be rewritten as:

$$
\begin{bmatrix} V'_m \\ F' \end{bmatrix} = \frac{1}{nlB} \begin{bmatrix} 1 & Z_e \\ Z_m & Z_m Z_e + n^2 l^2 B^2 \end{bmatrix} \begin{bmatrix} V' \\ I' \end{bmatrix}
$$

Finally, in Fig. 3.3(d) the impedance Z_e has been replaced by the resistance of the coil R_e in series with the inductance L_e of the coil. The mechanical admittance Y_m consists of the mechanical resistance R_m, the mass C_m of the moving part of the membrane and coil, and the mechanical compliance L_m of the suspension part of the membrane. If we know the magnitude of each parameter, we can now calculate the transfer characteristic and the dynamic behaviour. As one would expect, this behaviour depends on the electrical impedance with which the voice coil is terminated. The analogue of Fig. 3.3(d) also enables us to calculate the input and output impedances of this transducer. This information is necessary for matching the transducer correctly to the measurement object and to the remainder of the measurement system.

3.2.1 Mechanoelectric transducers

We will now discuss several operation principles of transducers for measuring mechanical quantities. We will confine ourselves to the more common quantities such as displacement, velocity, acceleration, force, etc.

Displacement transducers
Displacement transducers can measure either linear displacement (translation) or angular displacement (rotation). They can also be classified according to the *principle of transduction* on which they are based. We can distinguish between, for instance, resistive, capacitive, inductive and optical translation or rotation transducers. These mechanical transducers are also referred to as gauges or sensors.

Resistive displacement transducers
The potentiometric transducer is a popular type of displacement transducer. For measuring translation it is, in fact, no more than a sliding potentiometer. For measuring rotation, a rotating potentiometer can be used. Usually the potentiometers are wire-wound to achieve a better accuracy, temperature coefficient, etc. However, a limitation of wire-wound displacement transducers is their finite resolution. The maximal resolution $R = x/\Delta x$ is equal to the number of turns on the potentiometer body. Metal film potentiometers do not have this limitation. A disadvantage of all transducers of the potentiometric type is that mechanical wear and chemical corrosion can change the transfer characteristic over the lifetime of the transducer.

An example of the specifications of a linear displacement sensor, *in casu* a sliding NiCr wire wound potentiometer with 1000 turns is: range 25 cm; total resistance 300 Ω; non-linearity 10^{-3}; maximal resolution 10^3: temperature coefficient 2×10^{-4} K^{-1}; friction force 0.5 N.

It can be seen that a potentiometric displacement transducer becomes non-linear when it is loaded. The relative error which results from loading is zero when $x = 0$ or $x = 1$ and maximal when $x = 0.5$ (x is the relative position of the sliding contact (or wiper), with respect to the resistor body, so $0 \leq x \leq 1$).

When $x = 0.5$ the relative error is equal to $-R/4R_l$, where R is the total resistance of the potentiometer and R_l is the load resistance.

Another type of resistive displacement sensor makes use of the fact that the electrical resistance of a conductor depends on the dimensions of the conductor. The resistance R is a function of the cross sectional area of the conductor, its length l and its resistivity ρ:

$$R = R(A,l,\rho)$$

If the conductor is mechanically strained or compressed, the parameters A, l and ρ, and as a consequence R, will change. This enables one to measure very small displacements. Fig. 3.4(a) shows a piece of wire (strain gauge) which is elongated a distance Δl by applied tensile stresses. With a series expansion we can compute the sensitivity of this wire strain gauge:

$$R + \Delta R = R(A,l,\rho) + \left(\Delta A \frac{\partial}{\partial A} + \Delta l \frac{\partial}{\partial l} + \Delta\rho \frac{\partial}{\partial\rho}\right)R(A,l,\rho)$$
$$+ \frac{1}{2!}\left(\Delta A \frac{\partial}{\partial A} + \Delta l \frac{\partial}{\partial l} + \Delta\rho \frac{\partial}{\partial\rho}\right)^2 R(A,l,\rho) + \dots$$

If $\Delta A/A$, $\Delta l/l$ and $\Delta\rho/\rho < 1$ and the curvature at the point $R(A,l,\rho)$ is small, we can neglect the higher order terms. Subtracting $R = R(A,l,\rho)$ from both sides of the equation yields the following differential equation:

$$dR = \frac{\partial R}{\partial A} dA + \frac{\partial R}{\partial l} dl + \frac{\partial R}{\partial\rho} d\rho$$

With $\rho = \rho(l)$, $A = A(d)$ and $d = d(l)$ this results in:

$$dR = \frac{\partial R}{\partial A} \frac{\partial A}{\partial d} \frac{\partial d}{\partial l} dl + \frac{\partial R}{\partial l} dl + \frac{\partial R}{\partial\rho} \frac{\partial\rho}{\partial l} dl$$

Figure 3.4. (a) Freely suspended strain wire. (b) Metal foil strain gauge.

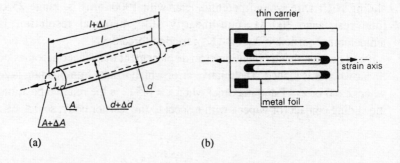

(a)   (b)

and therefore:

$$\frac{dR}{R} = \frac{dl}{l} \left\{ \frac{l}{R} \left(\frac{\partial R}{\partial A} \frac{\partial A}{\partial d} \frac{\partial d}{\partial l} + \frac{\partial R}{\partial l} + \frac{\partial R}{\partial \rho} \frac{\partial \rho}{\partial l} \right) \right\}$$

The factor in the braces is exactly the sensitivity S_l^R of the wire with respect to variations in length. We can calculate the magnitude of this sensitivity using the following equations:

$$R = \rho \frac{l}{A} \qquad \text{(electrical resistance)}$$

$$\frac{\partial \rho}{\partial l} = c \frac{\rho}{l} \qquad \text{(c depends on the material)}$$

$$A = \pi \frac{d^2}{4} \qquad \text{(circular cross section)}$$

$$\frac{\partial d}{\partial l} = -\mu \frac{d}{l} \qquad \text{(μ is Poisson's constant)}$$

We find:

$$S_l^R = 2\mu + 1 + c$$

If the volume remains constant upon elongation, $\mu = 0.5$ according to the last expression but one. For many materials $\mu \approx 0.3$. The resistivity of most metals does not depend on the strain; the constant c is very small. Consequently, the sensitivity is approximately 2 for most metals (chromium-nickel: 2.1–2.3; constantan: 2.0–2.1; chromel: 2.5, but manganin: 0.5 and nickel: −12).

One may assume for most metals that the volume of the conductor and the resistivity will not change with tensile or compressive stress (no piezoresistive effect). Therefore, for a *metallic strain gauge* we find:

$$\frac{dR}{R} = k \frac{dl}{l} \approx 2 \frac{dl}{l}$$

in which $k = S_l^R$ is the *strain gauge sensitivity factor*. Semiconductor materials generally exhibit a k value much larger than 2. This is due to the fact that no longer is $\partial p/\partial l \approx 0$. In these materials the piezoresistive effect dominates.

For an extrinsic semiconductor $\rho = 1/nq\mu$, in which n is the concentration, q is the charge and μ is the mobility of the charge carriers. Since $\mu = q\tau/m$ we find $\rho = m/nq^2\tau$, with m the effective mass and τ the relaxation time of the charge carriers. The effective mass is determined by the interaction between the charge carriers and the crystal lattice. When the material is stressed mechanically, the interaction and therefore the effective mass will change. A silicon strain gauge with a favourable crystal orientation may exhibit such a large piezoresistive effect that the sensitivity factor reaches values over 200.

Although semiconductor strain gauges may yield a larger sensitivity, they also exhibit a larger non-linearity and a larger temperature coefficient. Take, for instance, a metal foil strain gauge: material constantan; sensitivity factor $k = 2.00 \pm 1\%$; resistance $120\ \Omega \pm 1\%$; measurement range $10^{-6} \leq \Delta l/l \leq 10^{-2}$; non-linearity 10^{-3} for $\Delta l/l < 10^{-3}$; temperature coefficient $2 \times 10^{-5}\ \mathrm{K}^{-1}$; thermal expansion coefficient $1.4 \times 10^{-5}\ \mathrm{K}^{-1}$; thermoelectric voltage of a junction of constantan and copper $43\ \mu\mathrm{V/K}$. If we compare this to a semiconductor strain gauge the figures become: $k \approx 50\text{–}200$; non-linearity 10^{-2} for a measurement range of $\Delta l/l \leq 10^{-3}$; the temperature coefficient of k is approximately $10^{-3}\ \mathrm{K}^{-1}$.

As shown in Fig. 3.4(b), the meanders of a metal foil strain gauge are made extra wide where they turn, in order to reduce the sensitivity of the strain gauge to strain perpendicular to the main axis of operation. If, besides the magnitude of the strain, we also wish to measure the direction of the strain, a combination of strain gauges is used, arranged in a certain geometric pattern, for instance three strain gauges oriented at 120° angles with respect to one another. This arrangement is known as a *strain gauge rosette*.

For measuring the strain in a mechanical structure, the strain gauge is cemented to the structure in the direction of the expected stress. The characteristics of the cured glue and the carrier will give rise to *creepage effects*. If the strain is lasting, the metal foil or wire will slowly creep back to the original stressless state (stress relaxation). This effect is especially strong at higher temperatures. In addition, the glue and the carrier can also cause *hysteresis*. After being stressed, the metal foil or wire will not return immediately to its original state; it will appear as if there is still a small strain left. To keep the creepage and hysteresis effects small, the glue and carrier must be thin and hard and have a large Young's modulus.

The metal conductor of a strain gauge, the carrier and the material of the structure to which it is affixed must all have the same coefficient of thermal expansion. If the respective coefficients are not equal, an apparent strain will result from a change in temperature. There will also be an additional apparent strain due to the non-zero temperature coefficient of the resistance of conductor material used. Therefore, a second strain gauge is often used to

compensate these effects. The second strain gauge is located in such a way
that it is exposed to the same disturbances, but is not exposed to any stress (a
so-called passive or dummy strain gauge, see Fig. 3.5(a)). The strain gauges
are incorporated in a Wheatstone bridge, in the locations of R_1 and R_4 of the
bridge in Fig. 3.5(d). In this way most disturbances can be reduced
considerably.

Fig. 3.5(b) illustrates a method of measuring the bending of a cantilever
beam. Fig. 3.5(c) shows the measurement of the torsion of a shaft by means
of four strain gauges which are cemented to the shaft at 45° to the central
axis. Finally, Fig. 3.5(d) depicts a Wheatstone bridge configuration for
compensating disturbances. If the strain gauges R_1, R_2, R_3 and R_4 are
connected as indicated, the strain measurement of (a) will be insensitive to
temperature, the bending measurement of (b) will be insensitive to strain and
temperature and the torsion measurement of (c) will even be insensitive to
strain, bending, temperature and temperature gradients along the shaft. NB
In situations (a) and (b) the bridge resistors R_2 and R_3 are fixed resistors
whose values are chosen for maximum bridge sensitivity (see Section 3.3.3).

Figure 3.5. Measuring linear extension, bending and torsion with compensating
strain gauges in a Wheatstone bridge. (a) Compensated strain measurement.
(b) Compensated bending measurement. (c) Torque measurement.
(d) Measurement bridge.

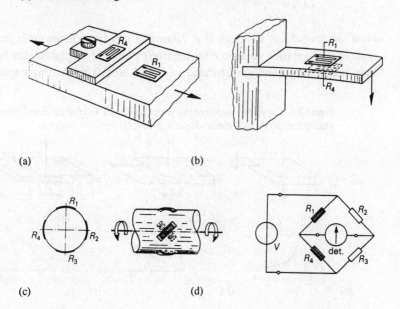

(a)   (b)

(c)   (d)

Capacitive displacement sensors

The capacitance C is a function of the distance d between the electrodes of a structure, the surface area A of the electrodes and the permittivity ε of the dielectric between the electrodes, so:

$$C = C(d,A,\varepsilon)$$

Evidently, there are three methods for realising a capacitive displacement sensor, i.e. by varying d, A or ε. These three methods are illustrated in Fig. 3.6. If we choose a capacitor with flat coplanar plates a distance x apart and ignore any edge fringe effects, the capacitance of this parallel plate transducer is given by:

$$C(x) = \frac{\varepsilon_0 A}{x}$$

This transducer is non-linear, however, and has the hyperbolic transfer characteristic of Fig. 3.6(a). This type of sensor is often used for measuring small incremental displacements without making contact with the measurement object. The transducer is usually linearised by using it in a balanced configuration. An example has already been given in Fig. 3.1(b).

If the surface area of the electrodes of the flat plate capacitor is varied, we find:

$$C(x) = \frac{\varepsilon_0 b x}{d}$$

Now the transducer is linear in x. Normally this type of sensor is implemented as a rotating capacitor for measuring angular displacements rather than the sliding version illustrated in Fig. 3.6(b). The rotating capacitor

Figure 3.6. Capacitive displacement sensors with: (a) variable distance between electrodes; (b) variable electrode surface area; (c) variable dielectric.

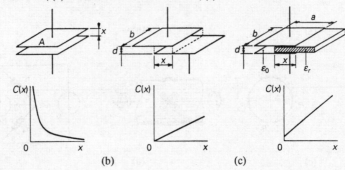

(a)   (b)   (c)

structure is also used as an output transducer for measuring electric voltages (capacitive voltmeter).

For a flat plate capacitor variation of the dielectric results in:

$$C(x) = C_0 \left\{ 1 + \frac{x}{a} (\varepsilon_r - 1) \right\}$$

in which $C_0 = C(0) = \varepsilon_0 ab/d$. This transducer is also linear. It is mostly used in the form of two concentric cylinders for measuring the level of a fluid in a tank. The non-conducting fluid forms the dielectric.

The force which a measurement object must exert on a capacitive transducer in order to move the electrodes is very small. We will consider this force $F(x)$ to be positive when it points in the direction in which x increases. Neglecting all losses (due to friction, resistance, etc.), the mechanical energy dE_m, supplied by the measurement object which causes an infinitesimally small displacement dx, plus the electrical energy dE_e supplied by the connected power supply V must be equal to the increase in field energy dE_f of the electrical field between the electrodes of the capacitor. The energy balance can be written as:

$$dE_m + dE_e = dE_f$$

in which:

$$dE_m = F(x)\, dx$$

$$dE_e = d(QV) = Q\, dV + V\, dQ$$

$$dE_f = d(\tfrac{1}{2} QV) = \tfrac{1}{2} Q\, dV + \tfrac{1}{2} V\, dQ$$

Since the supply voltage V across the capacitor is kept constant, it follows that $dV = 0$. Since $Q = VC(x)$, the Coulomb force is given by:

$$F(x) = -\frac{1}{2} V^2 \frac{dC(x)}{dx}$$

Therefore, the force on the transducer is positive in Fig. 3.6(a) and negative in Fig. 3.6(b) and Fig. 3.6(c). Thus, if the movable electrode had complete freedom of motion, it would assume a position in which the capacitance is maximal. If C is a linear function of x, the force F is independent of x.

Coulomb forces are extremely small. A linear capacitive displacement transducer of 100 pF with a measurement range of 1 cm and a supply voltage of 10 V requires a force of only 0.5 µN!

The capacity $C(x)$ in the three cases shown in Fig. 3.6 was calculated after neglecting the fringe or edge effects of the capacitors. These edge effects can be suppressed by using a shield electrode as indicated in Fig. 3.7. The displacement transducer C_t is equipped here with a shield electrode which is grounded and located such that the electric field inside the capacitor is uniform at the indicated locations (edges). If we measured only the capacitance of the moving electrode with respect to the fixed electrode, and did not also include that of the shield, the transducer would be free from disturbing edge effects. In Fig. 3.7(b) this is achieved by connecting the shield to the central connection of the transformer bridge. The voltage across the detector can be nulled by balancing the bridge with the variable capacitor C_n. Then the shield is at the same potential as ground. Since the capacitance between the not-grounded electrode of C_t and the shield is circuited in parallel with upper half of the secondary winding of the bridge transformer, it will not influence the balance condition of the bridge: $C_t = C_n$.

Often the capacitive displacement transducer is incorporated in a bridge network for measuring the capacitance and thus the displacement. Another possibility is to use the capacitor in an oscillator as one of the frequency determining components. The frequency is then a measure of the displacement. Capacitive transducers are robust and cheap. An example of an acoustic measurement transducer is the condenser microphone, which due to its high sensitivity, its wide dynamic range and bandwidth and its flat frequency response (30 kHz within 1 dB) is often used in sound-level measurements.

Inductive displacement transducers

It is not only possible to vary the self-inductance of a single coil as a function of the displacement to be measured, it is also possible to vary the mutual inductance between two coils as a function of the displacement. An obvious way of varying the inductance of a coil is to vary the effective number of turns. This principle is illustrated in Fig. 3.8(a). A different

Figure 3.7. Shield electrode construction for suppressing edge effects in capacitive transducers.

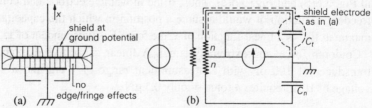

method is shown in Fig. 3.8(b) where the inductance is varied by varying the magnetic resistance (reluctance) of the yoke by means of an air gap of variable width.

If, for the first transducer, we disregard the inductance of the section of the coil which extends outside the yoke (which we may do if the permeability μ of the yoke is large and the air gap through which the coil passes is narrow) we find for both transducers:

$$L = \frac{\mu_0 n^2 A}{l_a + l_y/\mu_y}$$

In this expression n is the number of coil turns, A the cross sectional area of the yoke, l_a the length of the air gap, l_y the length of the path through the yoke and armature and μ_y the relative permeability of the ferromagnetic material of the yoke. The first principle in Fig 3.8(a) only varies n as a function of x, so:

$$\frac{dL(n)}{L(n)} = 2 \frac{dn}{n}$$

For the second principle in fig. 3.8(b), only l_a is varied. With $l_a = 2x$ the relation becomes hyperbolic in x:

$$L(x) = \frac{\mu_0 \mu_y n^2 A}{2x\mu_y + l_y}$$

This displacement transducer, based on a variable reluctance, is incorporated in a bridge network to obtain a linear transfer characteristic. Besides achieving linearity, the use of two of these variable reluctance transducers in a balanced configuration also nulls out the force between the yoke and armature. This magnetic force can be considerable, quite unlike the Coulomb force in capacitive transducers. From the energy balance we can deduce that this force is given to:

Figure 3.8. (a) Inductive displacement transducer based on varying the effective number of turns. (b) Inductive transducer with a variable magnetic resistance.

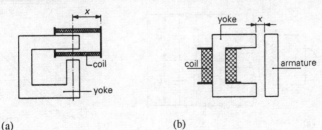

(a)   (b)

$$F(x) = -\frac{1}{2} I^2 \frac{dL(x)}{dx}$$

provided that the inductor is supplied with a constant supply current. Since $L(x)$ is not a linear function of x, the force F depends on x.

The magnitude of the inductance, which is the measure of the displacement, can be determined either with a bridge configuration, or by incorporating the inductor in an oscillator.

A *differential transformer* is an inductive transducer which makes use of the variation of the mutual induction between two coils. As the name indicates, this transducer is used in a balanced configuration (see Fig. 3.9). The coupling between the primary and secondary windings of the transformer depends on the position of the core. If, in Fig. 3.9(a) for instance, the core is moved up, the coupling between the primary and the upper secondary winding and thus the output voltage of this winding will increase. This AC voltage is converted into a DC voltage $\hat{V}_1$ with a peak detector. In Fig. 3.9(c) this voltage is plotted as a function of the position x of the core. The figure clearly shows that the relationship is highly non-linear. The output voltage of the lower secondary is plotted as $-\hat{V}_2$, in the same figure. The sum of these two voltages form the output voltage $V_o = \hat{V}_1 - \hat{V}_2$. For small excursions of x, around $x = 0$, the centre position of the core, the output voltage V_o is linear: a result of the non-linear characteristics of the two secondary windings being perfectly balanced. The differential sensitivity of this displacement transducer is proportional to the supply voltage applied to the primary winding.

Differential displacement transformers are available with measurement ranges from ± 1 mm to ± 25 cm. The non-linearity over the total range is approximately 2.5×10^{-3}. The operating frequency lies between 50 Hz and

Figure 3.9. (a) Differential displacement transformer with dual peak detectors. (b) Cross section of a differential transformer. (c) Transfer characteristic $V = V(x)$.

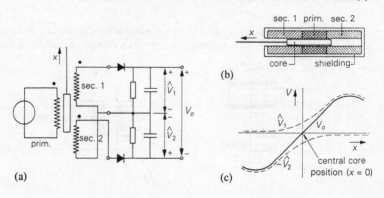

10 kHz. With the core in the central position and a supply voltage of 1 V, the sensitivity of a differential transformer with a measurement range of ± 1 mm is approximately 240 µV/µm; for a measurement range of ± 25 cm it is approximately 4 µV/µm.

Optical Displacement Transducers

Displacement can also be detected optically, by means of a encoder strip (translation) or a rotary encoder (rotation). Fig. 3.10(a) shows an optical displacement sensor which utilises an encoder strip consisting of rows of alternating transparent and opaque bars. The position of the strip is converted directly to a digital signal with a narrow beam of light and a number of light sensors. The digital code is determined by the position of the transparent and opaque bars on the strip.

It should be noted that the code used in an optical encoder is not usually a straightforward binary code, as in Fig. 3.10(a). A major disadvantage of such a binary code is the fact that, even with a small displacement, several bits may change at once. Due to the finite resolution of the optical detection system, it is possible for a state to occur in which a few (but not yet all) of the bits have changed. This (transitional) state may correspond to a totally different position of the encoder strip. Therefore, a code is usually chosen in which no more than one bit changes at a time, for instance, a Gray code.

Displacement sensors which use rotary or strip encoders have a fixed zero. Therefore, they are referred to as *absolute displacement transducers*. They are particularly well suited for connecting to a PC for digital position control.

Figure 3.10. (a) Optical displacement sensor employing an encoder strip.
(b) Enlargement of the distance between the grating lines employing the Moiré effect.

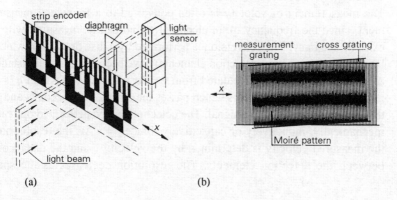

(a)   (b)

As an alternative to encoding the absolute position along the strip, it is also possible to use a single track of alternating transparent and opaque bars. Then a change in position can be determined by counting the number of light impulses. This method does not have a fixed zero and is therefore an *incremental displacement measurement*. The underlying measurement is an interval measurement. The resolution of the measurement is determined here by the smallest distinguishable distance between two adjacent grating lines. This distance is limited by the light sensor and is usually only 1 mm. If a higher resolution is desired, the Moiré effect can be used as illustrated in Fig. 3.10(b). A grating is placed in front of the original measurement grating at a small angle, causing a pattern of light and dark beams, which move vertically when the measurement grating is moved horizontally. When the grating is moved over the distance between two adjacent grating lines, the beams will travel precisely over a distance equal to their own width. The distance between the beams and the width of the beams is determined by the angle between the front grating and the measurement grating. When a light sensor is positioned behind this combination, it appears as if the distance between the grating lines has increased.

With this method displacement enlargements of up to 10^3 are obtainable. The resolution is of the order of 1 μm. This can be increased even further by using prisms and other optical means.

Velocity transducers
With transducers for measuring velocity, we must also distinguish between transducers for translational and angular velocity. If the angular velocity is nearly constant we also speak of a ppm measurement. In the following section we will discuss several methods and principles which are often used for converting velocity information into electrical information.

Velocity-to-frequency conversion
The measurement of velocity is often converted to a frequency measurement, since the frequency of an electrical signal can be measured with an extreme accuracy. The conversion is performed by a disc or strip on which a large number of marks (detection elements) have been put at equal distances Δx. The velocity can be calculated from $v = \Delta x \, n/t = \Delta x \, f$, in which n is the number of detection elements which passes the detector in t seconds and f is the frequency of the output signal. The detection can be performed optically, mechanically, inductively or capacitively. Since $f = v/\Delta x$, the resolution of the measured velocity is determined by the velocity v and the distance Δx between the detection elements. The resolution decreases as the speed

decreases and, therefore, for measuring low speeds the distance Δx must be very small.

Measurement of velocity by differentiation and integration

For a linear velocity $v(t)$:

$$v(t) = \frac{\mathrm{d}x(t)}{\mathrm{d}t}$$

where $x(t)$ is the linear displacement. For an angular velocity $\omega(t)$:

$$\omega(t) = \frac{\mathrm{d}\theta(t)}{\mathrm{d}t}$$

where $\theta(t)$ is the rotation angle.

From these two expressions it is evident that we can obtain a signal for the velocity of an object by calculating the derivative of the output signal of a displacement sensor. It is simple to design an electronic circuit for differentiating an electrical signal. However, there is one disadvantage associated with this: discontinuities in the output signal of the displacement sensor (due to its finite resolution) will cause large glitches in the velocity signal. In addition, high-frequency noise and other disturbances are amplified more by differentiation. This is because the transfer function of a differentiator increases linearly by 6 dB/octave with increasing frequency. Therefore, a signal for the velocity which is obtained by differentiation of a displacement signal is almost always afflicted with noise and interference.

Another possibility for obtaining the velocity is by integration of the linear acceleration $a(t)$:

$$v(t) = \int_0^t a(t)\,\mathrm{d}t + v(0)$$

or of an angular acceleration $\alpha(t)$:

$$\omega(t) = \int_0^t \alpha(t)\,\mathrm{d}t + \omega(0)$$

If we assume that at $t = 0$ the velocity was zero, it is sufficient to integrate the acceleration over the interval $(0,t)$. This can be performed by a relatively simple analogue electronic circuit. A disadvantage of this is that the output of any practical integrator will increase or decrease very slowly, even when the input is zero, due to leakage or droop. Therefore, integration in this manner is only feasible for relatively short intervals. Electronic integrators are often used for measuring the velocity of vibrating components of mechanical structures. Finally, it must be noted that the frequency characteristic of an integrator decreases linearly with increasing frequency at a rate of –6 dB/octave. Therefore, noise and disturbances are suppressed more, as the frequency of these undesirable effects increases.

Inductive velocity transducers

With inductive velocity transducers, the velocity of the measurement object is made to give rise to a change of magnetic flux Φ, which induces an electrical potential in a conductor. This induced voltage is a measure of the velocity of the measurement object. This will become apparent with the aid of Fig. 3.11(a). The induced voltage in turn i of the coil of this inductive velocity pick up is given by:

$$V_i = -\frac{\mathrm{d}\Phi_i}{\mathrm{d}t}$$

Therefore, for a total of n turns, the coil terminal voltage is:

$$V = -\sum_{i=1}^{n} \frac{\mathrm{d}\Phi_i}{\mathrm{d}t}$$

Since $\Phi_i = \Phi_i(x)$, in which x is the position of the magnet with respect to the centre of the coil, and $x = x(t)$ we find:

$$V = -\sum_{i=1}^{n} \frac{\mathrm{d}\Phi_i}{\mathrm{d}x} \frac{\mathrm{d}x}{\mathrm{d}t} = -v \sum_{i=1}^{n} \frac{\mathrm{d}\Phi_i}{\mathrm{d}x} = v\, k(x)$$

Thus, the output voltage V is proportional to the velocity v of the magnet for a given value of x. The sensitivity of the transducer is equal to k. Unfortunately, since $k = k(x)$, the transducer is non-linear. Therefore, again

Figure 3.11. Inductive velocity transducers. (a) A magneto-dynamic velocity transducer. (b) An induction flow rate transducer.

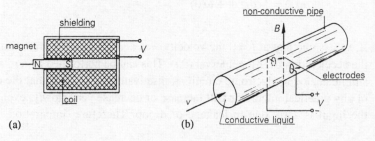

(a)   (b)

this type of transducer is often used in a balanced configuration. Since the magnet is moving here the velocity sensor in Fig. 3.11(a) is referred to as a magneto-dynamic transducer.

Another example of an inductive velocity transducer was given in Fig. 3.3. Here, the magnet is stationary and the coil moves, and therefore this transducer is referred to as an electro-dynamic transducer. Fig. 3.11(b) shows an example of a flow rate measurement. A conductive liquid flows through a pipe between two electrodes and is exposed to a strong magnetic field with induction B. This field induces an electrical potential difference at the electrodes which can be calculated as follows: Assuming the induction B and the flow rate are constant throughout the entire pipe cross section and also assuming that $\mathbf{B} \perp \mathbf{v}$, then the change of flux $\Delta\Phi$ per second in a column of liquid between the two electrodes will amount to $B\Delta A = Blv$, were l is the distance between the electrodes. Thus, the induced voltage with the polarity indicated in Fig. 3.11(b) is given by:

$$V = Blv$$

The sensitivity of an inductive flow rate transducer is relatively low. For instance, in a strong magnetic field of 0.1 T, a pipe with a diameter of 5 cm and a flow rate of 1 m/s will only produce an output voltage of 5 mV. In principle, the transducer is linear, but due to variations in the flow rate over the cross section of the pipe, caused by the viscosity of the fluid, this is not always the case. It can be shown that the output voltage V is a linear measure of the cross sectional average of the flow rate: $v_a = Q/\pi r^2$, in which Q is the total volume of liquid passing through a cross section πr^2 per second.

Acceleration transducers

Transducers for measuring acceleration rely on the measurement of the force F required to give a known mass (the seismic mass m) the same acceleration (a) as the measurement object. From the force and the mass the acceleration is determined: $a = F/m$. However, attaching this mass (and the additional mass of the transducer housing) to the measurement object may influence the measured acceleration. Therefore, the extra mass must be kept to a minimum, especially when the measurement object is highly elastic or has a low mass. We will not deal with acceleration sensors separately, as the necessary methods and principles of transduction will be discussed in the following subsection on force transducers.

Force transducers

In order to measure a mechanical force, an elastic body can be subjected to this force. The resulting deformation or change in dimensions of this body can subsequently be measured with a displacement sensor. The shape and material of the elastic body in such a 'dynamometer' must be chosen so as to satisfy Hooke's law over a large range without plastic deformation. For a cylindrical rod Hooke's law gives:

$$\sigma = \frac{F}{A} = E\frac{\Delta x}{x}$$

Here σ is the stress in the rod, F the force exerted on the rod, A the cross sectional area of the rod and $\Delta x/x$ the relative extension (or strain) of the rod. E is the modulus of elasticity of the material. Many different types of spring are used to obtain a large linear range for strain and stress (see Fig. 3.12). Pressure is often measured by means of a hollow spring which straightens (unwinds) under pressure. Fig. 3.12(b) and (c) show these so-called Bourdon pressure gauges. The displacement x or rotation θ of these gauges is usually measured by means of strain gauges or differential displacement transformers.

Piezoelectric force transducers

Some materials exhibit an electrical polarisation that changes with any mechanical deformation of the material. When a force (or pressure) is exerted on a small body of such a material, a difference in electrical charge occurs between two opposite surfaces of the body. This phenomenon is called the *piezoelectric effect* (Gr. *piezein* = to press).

This effect is exhibited by materials with a crystal lattice which lacks a centre of symmetry, such as quartz. Any external mechanical strain will cause a change in the dipole moment of the crystal. This change is not the same in all directions in an asymmetrical crystal, causing a macroscopic electrical polarisation which results in a difference in charge at the crystal surfaces. This effect can also occur in materials which do have a centre of symmetry, but only if the symmetry is disturbed by a strong electrical field due to the spontaneous polarisation of the material. An example of this last class of materials (the so-called ferroelectrics) is barium titanate.

The reverse effect can also occur: when a voltage is applied to a piece of piezoelectric material, it will deform.

The piezoelectric effect can be used for a pressure transducer in the manner illustrated in Fig. 3.13(a). The charge Q building up at the surfaces of the crystal is proportional to the applied force F where the pressure $P = AF$ and A is the area of the pressure plates. The *charge sensitivity S_q* of the piezoelectric force transducer is defined as:

$$S_q = \frac{Q}{F}$$

This sensitivity depends on the material of the crystal and on its orientation, but it is independent of the dimensions of the crystal. The *voltage sensitivity S_v* of the transducer is defined as:

Figure 3.12. A number of different elastic spring bodies for converting a force F or a pressure P into a linear displacement x or a rotation θ: (a) ring type spring; (b) Bourdon helix spring (c) Bourdon spiral spring; (d) membrane spring.

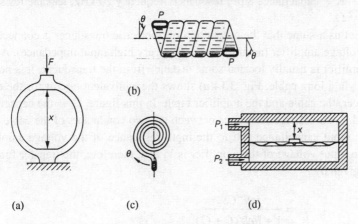

(a)   (c)   (d)

Figure 3.13. (a) Piezoelectric pressure transducer. (b) The equivalent electrical circuit.

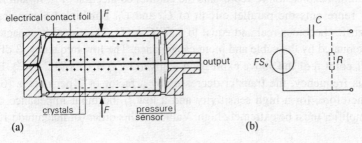

(a)   (b)

$$S_v = \frac{V}{F}$$

and since for a capacitor $Q = CV$ it follows that:

$$S_q = CS_v$$

in which C is the electrical capacitance of the transducer. Evidently, the voltage sensitivity of a piezoelectric transducer depends on its dimensions.

Fig. 3.13(b) shows the equivalent electrical circuit of such a piezoelectric transducer. The resistor R between the two terminals is the leakage resistance which is usually very high. This makes it impossible to measure static forces with this type of transducer. A static force would cause a DC voltage at the output. However, such a DC voltage bleeds off rapidly due to the charge leakage via any moisture film, surface contamination and non-infinite bulk resistance.

Some typical specifications of a practical piezoelectric pressure gauge are: material quartz; measurement range 0–5000 N/cm^2; non-linearity 1%; charge sensitivity 3 pC cm^2/N; temperature coefficient of the sensitivity $10^{-3}\,K^{-1}$; capacitance 8 pF; resonance frequency 20 kHz; leakage resistance $10^{12}\,\Omega$.

Let us assume that the output of a piezoelectric transducer is connected to a voltage amplifier (an amplifier with a very high input impedance). As this amplifier is usually located some distance from the transducer, it is hooked up via a long cable. Fig. 3.14(a) shows the equivalent circuit for the transducer, the cable and the amplifier input. In this figure, C_c is the capacitance and R_c the leakage resistance between the two conductors of the cable. C_i is the input capacitance and R_i the input resistance of the voltage amplifier. The input voltage of the amplifier is V_i and, therefore, the transfer function is given by:

$$\frac{V_i}{F} = \frac{j\omega R'CS_v}{1 + j\omega R'(C + C')}$$

In this expression R' represents the parallel connection of R, R_c and R_i and C' represents the parallel circuit of C_c and C_i. For high frequencies, the transfer becomes real and equal to $S_vC/(C + C')$. The signal is capacitively attenuated by the cable and input capacitance. The low frequency -3 dB cut-off point f_l of the above expression is given by $f_l = 1/2\pi R'(C + C')$. Below this frequency, the transfer decreases by a factor of 2 per octave (6 dB). Therefore, for a high sensitivity and a low f_l the input impedance of the amplifier must be extremely high. Values of this order (of magnitude $10^{14}\,\Omega$

in parallel with 1 pF) can be obtained with a special kind of measurement amplifier: the *electrometer amplifier*.

It is possible to avoid the problems imposed by the cable capacitance and the amplifier impedance by considering the charge as a measure for the force, rather than the voltage. The output signal of the transducer is then supplied to a *charge amplifier*, as shown in principle in Fig. 3.14(b). If the gain A_0 of the operational amplifier used is very large, the input voltage will be negligible, for any finite given output voltage V_o. This means that the voltage across the cable and across input impedance of the operational amplifier becomes approximately zero. Neglecting these impedances therefore, we see that the output charge of the transducer will flow entirely into the feedback impedance consisting of C_o in parallel with R_o. Therefore, the transfer function is equal to:

$$\frac{V_o}{F} = -\frac{j\omega C R_o S_v}{1 + j\omega C_o R_o} = -\frac{j\omega R_o S_q}{1 + j\omega C_o R_o}$$

Provided that R_o is large enough for $\omega R_o C_o$ to be larger than unity, the transfer will be real and is given by:

$$\frac{V_o}{F} = -\frac{S_q}{C_o}$$

An operational amplifier will always draw or supply a small input current (necessary for biasing the input transistors of the amplifier). Therefore, if R_o were omitted, the charge amplifier would integrate this current until finally the output would saturate. However, R_o must be large for the charge amplifier to operate correctly at low frequencies. As a consequence, the operational amplifier must have extremely small input bias currents (and use, for instance, field-effect transistor (FET) devices in the input stage).

The output voltage V_o of the charge amplifier is proportional to the generated charge Q, because $V_o = -FS_q/C_o = -Q/C_o$. The low impedance voltage V_o can subsequently be used for further transmission processing.

Figure 3.14. (a) Read-out of the output voltage of a piezoelectric force transducer interconnected by a cable. (b) The principle of change amplification.

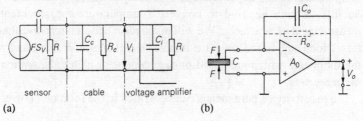

(a)   (b)

3.2.2 Thermoelectric transducers

In every temperature measurement, energy will be extracted from or supplied to the measurement object. The latter case occurs when the measured temperature is lower than that of the transducer which tends to be at ambient temperature. This heat exchange occurs with every measurement. The measurement object will be affected by this heat exchange and this gives rise to measurement errors. The transfer of energy from measurement object to temperature sensor, or *vice versa*, occurs by means of conduction, convection or radiation. The transfer of heat by stationary matter is called *heat conduction*. For a contact temperature sensor, heat conduction will be the primary form of energy transfer. The transfer of thermal energy by moving mass is called *heat convection*. This is the most important method of energy transfer when a temperature sensor is placed in a gas flow. Finally, the transfer of heat by infra-red electromagnetic radiation is referred to as *heat radiation*. No matter is involved in this process. Radiation thermometry relies on this type of energy transfer.

In this chapter we will discuss four essentially different conversion principles in use for thermoelectric transducers. They are all based on the temperature dependence of a variable or parameter of a measurement system. The first class relies on the temperature dependence of an electrical component; for instance, the resistance of a component or the band gap of silicon in a bipolar transistor. The second class measures the contact potential difference of two metals giving rise to so-called thermocouples. The third class measures the heat radiation, which is emitted by a body above $T = 0$ K (pyrometers). Finally, the fourth class measures temperature indirectly. The temperature first influences a mechanical quantity (for instance, the resonance frequency of a crystal), which in turn influences an electrical quantity (the oscillator frequency of an oscillator). This is used in a quartz crystal thermometer. Table 3.1 gives a short summary of the advantages and disadvantages of these classes.

Resistive temperature sensors

The electrical resistance of any material depends to a certain extent on the temperature. If this dependence is accurately known and sufficiently reproducible, it may be used to convert a temperature measurement into a measurement of resistance. We will distinguish between resistive thermometers which are based on pure metals, *metal thermometers*, and resistive thermometers which are based on semiconductor materials, *semiconductor thermometers*.

The resistivity of pure metals can be written in the form of a power series:

$$R(T) = R(T_0)[1 + \alpha(T - T_0) + \beta(T - T_0)^2 + \gamma(T - T_0)^3 + ...]$$

in which $R(T)$ is the resistance of the sensor at temperature T and $R(T_0)$ is the resistance at a certain reference temperature T_0. If the temperature range is not too large, the first two terms of the expression will suffice; the sensor is approximately linear.

The most frequently used metals are platinum and nickel. At $T_0 = 273$ K, $\alpha_{Pt} = 3.85 \times 10^{-3}$ K^{-1} and $\alpha_{Ni} = 6.17 \times 10^{-3}$ K^{-1}. The measurement range of a platinum sensor runs from 70 K to 1000 K and that of a nickel sensor from 200 K to 500 K. The second order term for platinum is $\beta_{Pt} = -5.83 \times 10^{-7}$ K^{-2} and the third order term $\gamma_{Pt} = -3.14 \times 10^{-12}$ K^{-3} at 273 K.

The resistance of a pure metal whose lattice has no impurities or dislocations will exhibit a positive temperature coefficient α. The resistance is caused by the interaction between the free conduction electrons and the vibrating atoms of the crystal lattice. The amplitude of the lattice vibrations will increase as the temperature increases, resulting in a decrease in the mean path length of the electrons and a reduction in the mean time τ (relaxation time) between collisions. The relaxation time τ can be proven to be proportional to the reciprocal of the absolute temperature and therefore the resistance is proportional to absolute temperature.

Table 3.1. *Comparison of various methods for measuring temperature.*

	Resistive Sensor	Thermistor	IC Sensor	Thermo-couple	Radiation Sensor	Quartz Sensor
Advantages	stable accurate	large effect fast two-wire measurement	cheapest linear large voltage	passive transducer simple robust cheap large temperature range	no load measures at a distance	most stable most accurate most linear
Disadvantages	four-wire measurement non-linear more expensive slow produces heat	less stable very non-linear produces heat small temperature range fragile	slow produces heat requires power source $T < 150$ °C	most non-linear small voltage requires temperature reference not very stable	inaccurate not stable more expensive requires temperature reference	most expensive

In both intrinsic and extrinsic semiconductors this effect is overshadowed by a much stronger effect: the number of free charge carriers depends on the absolute temperature. The higher the temperature, the larger the number of electrons which cross the band gap from the valence band into the conduction band (intrinsic semiconductors) or the larger the number of activated donor and acceptor atoms (extrinsic semiconductors). The number of free charge carriers increases according to:

$$n = n_0 \, e^{-E_g/2kT}$$

in which E_g is the energy required for crossing the band gap and k is Boltzmann's constant. Thus, the resistance of a semiconductor will decrease as the temperature increases; a semiconductor has a Negative Temperature Coefficient (NTC-resistor).

The resistance of semiconductors can be expressed as:

$$R(T) = A \, e^{B/T}$$

The coefficients A and B also depend on the temperature and therefore a more accurate expression is given by:

$$R(T) = R(T_0) \, e^{B(1/T - 1/T_0)}$$

Obviously, a semiconductor sensor is highly non-linear. The temperature coefficient $\alpha(T)$ is given by:

$$\alpha(T) = \frac{1}{R(T)} \frac{dR(T)}{dT} = -\frac{B}{T^2}$$

For practical semiconductor sensors, the value of the coefficient B lies between 2700 K and 5400 K at a temperature of 300 K. Thus, at 300 K, the temperature coefficient ranges from -3×10^{-2} K^{-1} to -6×10^{-2} K^{-1}.

At 300 K, a semiconductor sensor is an order of magnitude more sensitive than a metal sensor. Such a semiconductor temperature-sensing resistor is often referred to as a *thermistor*.

Several practical remarks must be made on the use of metal sensors. The variation of resistance per kelvin is rather small and, therefore, it is necessary to apply the measures of Section 2.3.3.3 to avoid disturbances. Twisted conductors and screened cables are needed. Most metal sensors have a resistance of approximately 100 Ω and are made of platinum. The construction of the carrier is such that the mechanical stress due to thermal expansion of the bobbin on which the resistance sensor wire is wound is minimised. Finally, the sensor will be heated slightly by the dissipation of the sense current in the sensor resistance (this self-heating is approximately 0.5 K in stationary air and 0.1 K in an air flow of 1 m/s).

The basic principle of operation of the temperature sensors discussed so far utilises a temperature dependent resistance. The resistance change is usually measured by means of a bridge configuration, often located at some distance away from the actual temperature sensor. The connecting cable will have its own resistance. Worse still, the cable resistance will also depend on the temperature. As a consequence, the ambient temperature will contribute to the measurement result; the measurement is sensitive to environment disturbances. The sensitivity to ambient temperature can be suppressed by adding one or two extra leads to the measurement system, thus obtaining a *three-wire circuit* or a *four-wire circuit*. Fig. 3.15(a) shows that in a three-wire circuit one makes use of the fact that two of the leads have identical resistance, *in casu* $R_{a1} = R_{a2}$. The four-wire circuit does not have any such requirements but needs one more lead.

IC temperature sensors

An alternative temperature sensor can be found in the bipolar transistor. Such a sensor makes use of the fundamental band gap voltage of silicon, which depends on the temperature. The sensor is implemented as follows. Two bipolar transistors located close together, on the same integrated circuit (IC), are biased to different collector currents. If the ratio of the (collector) *current densities* (the transistors may have different areas) is equal to r, the difference between the base-emitter voltages of the two transistors is equal

Figure 3.15. (a) Three-wire circuit; if $R_{a1} = R_{a2}$ and $R_2 = R_3$ then $R(T) = R_4$. (b) Four-wire circuit; if $R_i \gg R(T)$ then $R(T) = V/I$.

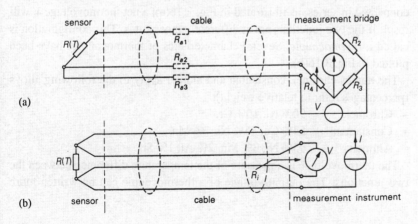

to $(kT/q) \ln r$. Here k is Boltzmann's constant, T is the absolute temperature and q is the charge of an electron. This base-emitter voltage difference is a linear measure of the absolute temperature. Additional electronic circuits amplify this voltage to provide a practical output value.

Typical IC sensor specifications are: temperature range from –55 °C to 150 °C; non-linearity over the entire range of approximately 0.3 K; sensitivity 10 mV/K (voltage output) or 1 µA/K (current output); instability over 1000 h of operation ± 0.08 K; dissipation 1.5 mW–3 mW. The thermal resistance of the sensor to ambient is roughly 200–400 K/W in stationary air. The temperature rise due to self-heating is then about 0.3 to 1.2 K. The thermal resistance and, therefore, also the self-heating can be reduced by about a factor of 4 when the effective surface area of the sensor is increased by a heat sink. A disadvantage of this solution is that the sensor then responds more slowly, due to the increased thermal mass. A time constant of $\tau = 80$ s for a sensor including the casing in stationary air may increase to $\tau = 120$ s when a heat sink is attached. In some cases it is also possible to reduce the self-heating by positioning the sensor in stirred air. At an air flow rate of 3 m/s, the thermal resistance is decreased by a factor of 5. If these sensors are used in fluids, especially when the fluid is flowing, the self-heating becomes negligible.

Thermocouples

When two different metals are brought into atomic contact with each other an electrical potential difference is generated. This so-called junction potential depends only on the nature of the two metals and on the absolute temperature. The surface area of the junction has no influence. For many combinations of metals the junction potential difference is approximately linearly proportional to the absolute temperature of the junction, provided that the temperature range is not too large. When two junctions are connected in series, as illustrated in Fig. 3.16(a), a net thermovoltage V will result if the two junctions are at different temperatures. This configuration is called a *thermocouple*. Several characteristics of thermocouples have been plotted in Fig. 3.16(b).

The names chromel, constantan and alumel apply to the following alloys (percentages refer to relative weight):
-   Chromel: 90% Ni, 10% Cr.
-   Constantan: 54% Cu, 45% Ni, 1% Mn.
-   Alumel: 95% Ni, 3% Mn, 2% Al, 1% Si.

The thermovoltage is a measure of the temperature difference between the two junctions. The output voltage of a thermocouple can be written more accurately as a power series of the temperature difference $T - T_0$, with T_0 a certain calibration temperature than it can if it is assumed to be proportional to temperature: i.e.

$$V = a_1(T - T_0) + a_2(T - T_0)^2 + \dots + a_n(T - T_0)^n$$

As n increases, this expression will describe the behaviour of a given thermocouple more and more accurately. Every thermocouple (so every combination of two metals) is characterised by its own series of *temperature independent* coefficients a_i ($i = 1, \dots, n$). An inaccuracy of $\pm 1\%$ requires roughly eight coefficients ($n = 8$) for most materials. The coefficient a_1 is referred to as the *Seebeck coefficient*. Fig. 3.16(c) shows how this coefficient would depend on the temperature if a thermocouple were to be described by only one single coefficient, a_1.

If we are dealing with a large temperature range, we must use more than one coefficient for reasons of accuracy. This is easy when a computer is available. We can then enter a linear piecewise approximation of the employed thermocouple into a computer. This is especially worthwhile when a large number of temperatures are measured, as for instance is the case in an oil refinery. In this kind of application the thermocouples provide a relatively cheap and robust, but non-linear means of measuring temperature. The central computer can then linearise the read-out of each sensor.

For a small range around $T_2 = 0$ °C the following values may be used for the Seebeck coefficient a_1:

Figure 3.16. (a) Thermocouple consisting of two junctions of dissimilar metals which are held at different temperatures T_1 and T_2. (b) Transfer characteristics of a number of commonly used thermocouples (the letters indicate the various combinations of metals) E: chromel–constantan; J: iron–constantan; K: chromel–alumel; T: copper–constantan; R: (87% Pt, 13% Rh)–platinum; S: (90% Pt, 10% Rh)–platinum. (c) Seebeck coefficient of these thermocouples as a function of the temperature T_1 for a given reference temperature $T_2 = 0$ °C.

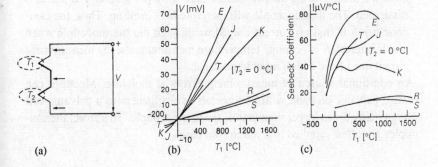

type E:	61 μV/K	type K:	40 μV/K
type J:	52 μV/K	type R:	6 μV/K
type T:	41 μV/K	type S:	6 μV/K

There are four physical effects that contribute to the output voltage of the thermocouple

- *Seebeck effect*. This is the desired effect which is measured when no current is allowed to flow through the thermocouple. It arises from the temperature dependence of the junction potential difference. This junction potential difference originates from the difference in Fermi levels of two dissimilar metals. The higher the temperature, the larger the number of electrons with a higher energy level than the Fermi level. This causes the junction potential difference to become temperature dependent.

- *Peltier effect*. When a current flows through a junction of two dissimilar metals, the temperature of the junction will change. Depending on the direction of the current, the junction will become either warmer or cooler than ambient. This effect is caused by the fact that with every electrical conduction process there is also transportation of heat. In a metal, thermal conduction as well as electrical conduction is caused by free electrons. The Peltier effect is undesirable in a thermocouple since it gives rise to a temperature error.

- *Thomson effect*. If a current is flowing through a uniform metal conductor in the direction in which there is a negative temperature gradient, thermoelectric heat will be generated. If the direction of the current is reversed, heat will be extracted from the conductor. This reversible effect also originates from the fact the electrical conduction process in a metal is accompanied by the transfer of heat and, inversely, heat conduction is accompanied by electrical conduction. This effect also gives rise to errors.

- *Joule heat*. With the last two effects, we assumed that no heat was generated by the dissipation of electrical energy in the electrical resistance of the metals. If the total resistance is R, then, per second, I^2R joules is dissipated. The thermocouple will therefore heat itself up. Thus, the clear conclusion is that no current may flow through the thermocouple when one is interested in accurate temperature measurements; the measurement circuit must have a high input impedance.

An additional source of measurement errors is moisture. Moisture can create a galvanic element with both metals, which generates a galvanic cell voltage many times larger than that of the thermocouple. Therefore, thermocouples are often supplied in a waterproof case.

If we wish to measure the absolute temperature as opposed to a temperature *difference* by means of a thermocouple, we must hold one of the junctions at a fixed known reference temperature. This can be achieved by controlling the temperature of one of the junctions with a thermostat. It is also possible to *compensate* the temperature of the *reference junction*, as indicated in Fig. 3.17. The temperature of the reference junction is measured here by a resistance sensor $R(T)$, which is connected to a bridge network. The output voltage of the bridge is connected in series with the thermocouple such that it compensates for the temperature of the reference junction.

Of course, the temperature sensitivity of the reference junction must oppose that of the bridge network. Often though, the absolute temperature of the reference junction is measured, especially when a large number of temperatures need to be measured simultaneously. All reference junctions are then located together on one reference block. The temperature of this block is measured, for instance, with an IC sensor. This temperature is added to the measured temperature differences of the individual thermocouples. The temperature measured by the active junction AB is usually located some distance away from the rest of the circuitry. The metal of the thermocouple junction is usually too expensive to use for long interconnections and, therefore, two wires $A'B'$ of a cheaper metal are used for the interconnections. Provided that these two wires A' and B' have the same thermoelectric characteristics as the two wires A and B of the measurement junction, no error will be introduced. It can be seen that when two (other) metal conductors are connected in series with the two junctions of the thermocouple, as in Fig. 3.17, the new junctions will not contribute a net potential difference, as long as the new junctions are kept at the same temperature (isothermal). The extra junction potentials will cancel each other. Temperature differences between these new junctions will cause

Figure 3.17. Extension of the measurement junction AB by means of a compensation cable $A'B'$ and the compensation of the reference junction temperature by a bridge network.

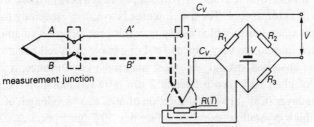

measurement errors. If the cable which connects the two junctions of the thermocouple has the same thermoelectric characteristics it is referred to as *compensation cable*.

When using thermocouples for temperature measurements, we must bear in mind a number of most prevalent sources of errors. Over the years, several alloys have been developed for thermocouples, each with its own specific application. These alloys are readily available as wire with a high grade of purity, uniformity, etc. so that the standardised coefficients a_1, ..., a_n of the alloy can be used. A thermojunction can be made simply by soldering, provided that there is no temperature gradient across the solder. In fact, soldering introduces a third metal between the two original thermocouple materials but is held at the same (measurement) temperature and, therefore, contributes no net thermovoltage. For applications at high temperatures a thermocouple must be spot-welded (by discharging a capacitor across the junction as it is clamped together). The behaviour of a thermocouple may drift due to the diffusion of gases in the metal, especially at high temperatures. Also, when a wire of the thermocouple crosses a large temperature gradient, it is possible that non-uniformities and impurities in this wire cause their own thermovoltages. Consequently, thermocouple wire is expensive and therefore thin. This reduces the thermal response time, but also increases the series resistance. Therefore, a current through the thermocouple, due to leakage resistance, will introduce the influence of the other effects discussed above. Finally, galvanic voltages due to moisture and corrosion can become several hundred times larger than the Seebeck voltage!

Radiation thermometers

A radiation thermometer absorbs a fraction of the infra-red radiation emitted by the measurement object. A radiation thermometer for high temperatures is usually called a *pyrometer* (Gr. *pyr* = fire). The radiation is usually focused on the actual thermal detector by means of a concave mirror (as illustrated in Fig. 3.18(a)). If the temperature of the measurement object is lower than that of the detector, it will supply heat energy to the object, causing the detector to cool. The use of lenses is avoided, especially for low temperatures, since lenses which easily transmit heat radiation (diathermanous infra-red lenses) are very expensive. Long wave infra-red radiation, in particular, is absorbed strongly by most materials. For instance, a 2 mm thick piece of glass will absorb 50% of 2 µm infra-red radiation. A 2 mm thick fluorite layer (CaF_2) has an absorption of 50% at a wavelength of 7 µm and 2 mm thick crystalline caesium iodide has 50% absorption at 70 µm.

These alternative materials are difficult to process and often are not resistant to moisture, etc.

As shown in Fig. 3.18(a), the distance from the measurement object to the pyrometer does not matter, since object A and object C both transfer the same heat radiation to the pyrometer when their surface areas completely fill the opening angle θ. The radiation of a point source decreases with the square of the distance, but the measured area increases with the square of the distance, so the net effect is zero if the viewing angle of the pyrometer is completely covered by an (isothermal) object.

If an object radiates uniformly in all directions (a so-called Lambert radiator), the radiation which the detector receives will not change, even when the measured surface is at an angle α (provided $\alpha \neq 0°$ or $180°$) with respect to the axis of the pyrometer.

The radiation energy emitted by a measurement object at a given surface temperature is usually less than that of a black body radiator at the same temperature; the measurement object has an emission coefficient lower than unity. (A black body radiator has an emission coefficient of 1 and a perfect reflector has an emission coefficient of 0.) The emission coefficient of a surface is equal to the absorption coefficient. Both can depend on the wavelength of the radiation. Some objects which reflect visible light very well may have a large emission coefficient in the infra-red region and *vice versa*.

The emission coefficient ε of various materials in the infra-red region is $\varepsilon \approx 0.03–0.05$ for polished metallic surfaces, $\varepsilon \approx 0.9$ for layers of varnish and enamel, $\varepsilon \approx 0.92$ for colloidal graphite and $\varepsilon \approx 0.96$ for water and ice. An object is often coated with a thin layer of colloidal graphite in order to obtain a large ε. Sometimes a hole is drilled in the object. When the

Figure 3.18. (a) The optical arrangement of a infra-red radiation thermometer for matching a thermal detector to the measurement object. (b) Simplified representation of the spectral sensitivity $S(\lambda)$ of quantum and thermal detectors.

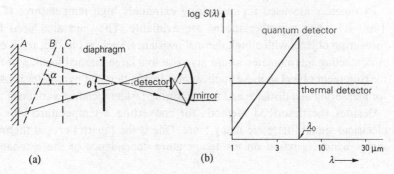

(a)   (b)

diameter-to-depth ratio is smaller than 1/6, the hole will behave like a virtually ideal black body radiator.

When the emission coefficient ε is not equal to 1, the measurement performed by a pyrometer will be in error. The power density P of the radiation satisfies the Stefan–Boltzmann law $P = \varepsilon c T^4$, in which c is a constant ($c = 5.7 \times 10^{-8}\,\mathrm{W\,m^{-2}\,K^{-4}}$) and T is the absolute temperature. Therefore, the error factor in the temperature measurement due to ε being less than 1 is $\varepsilon^{1/4}$.

Two different types of detector are used in pyrometers: thermal and quantum detectors. A *thermal detector*, or *absorption detector*, converts the incident radiation into a temperature rise. This temperature rise can then be measured with one of the principles discussed above; for instance, with a thermistor or by means of a number of thermocouples connected in series (a thermopile). The advantage of thermal detectors is their constant sensitivity over a large range of wavelengths (see Fig. 3.18(b)). However, their response is slow due to their relatively large thermal mass. Typical instrument reading times will lie between 1 and 10 seconds.

Other pyrometers are based on *quantum detectors*. This principle of operation relies on the electrons of the material being excited by infra-red radiation. This can only occur when the quantum energy E of the radiation quanta is equal to or larger than a certain threshold energy E_0. This threshold energy corresponds to the transition of electrons to a higher energy state. With:

$$E_0 = h f_0 = \frac{hc}{\lambda_0}$$

this threshold energy is related to infra-red radiation above a certain minimal frequency f_0 or below a certain maximal wavelength λ_0. Fig. 3.18(b) shows *in abstracto* the spectral sensitivity of this kind of quantum detector. Photodiodes are used as quantum detectors for near infra-red radiation; photoresistors are used in the far infra-red region. Quantum detectors respond quickly, but only measure radiation in a limited band of wavelengths.

Pyrometers are used for measuring extremely high temperatures ($T > 1000\,\mathrm{K}$) when no other means are available. They are also used for measuring objects with a high thermal resistance, such as plastics, rock, etc. A contacting thermometer would give rise to a large thermal loading of the measurement object in these applications. In addition, the pyrometer is used for measuring at a distance and for measuring fast temperature variations.

Besides the described methods for converting a temperature into an electrical signal, there are many more. One is the (quartz) crystal thermometer which is based on the temperature dependence of the resonance frequency of a piezoelectric crystal. The crystal is used to determine the frequency of an oscillator accurately. This frequency can be determined by the enumeration method, resulting in a small inaccuracy of only 0.01 K and a very high resolution of 10^4 in the range from –80 °C to 250 °C. These excellent specifications are realised by cutting the quartz crystal very accurately with respect to the crystal orientation axes. This is done in such a way that the temperature sensitivity of the resonance frequency is maximal and as stable as possible. A calibration curve of temperature versus frequency is recorded and built into the instrument (PROM) to increase the accuracy even further.

3.2.3. Magnetoelectric transducers

The induction of a magnetic field can be measured with transducers referred to as magnetometers or magnetic field sensors.

The magnetic induction B is expressed in teslas (T). Sometimes, it is also referred to as the magnetic flux density. A unit which is consistent with this terminology is the weber per square metre with, of course, $1 \, T = 1 \, Wb/m^2$.

Sometimes a rotating coil (with area A , n turns and angular frequency ω) is used for measuring the induction B of a static magnetic field. Assuming the coil is so small that the induction is constant across the surface area of the coil and B_n is the component of B perpendicular to the axis of rotation, the flux Φ through the coil is equal to $\Phi = B_n A \sin \theta(t)$, where θ is the instantaneous angle between the coil and B_n. With $\theta(t) = \omega t$, the induced AC voltage is given by:

$$ V = -n \frac{d\Phi}{dt} = -nB_n A \omega \cos \omega t $$

We can determine B_n from this expression. *Induction sensors* always require a changing flux. The change in magnetic flux can obviously also arise from an alternating magnetic field in a static coil.

Another type of magnetometer is based on the influence of a magnetic field on the electrical resistance of a material. As early as 1856 W. Thomson (Lord Kelvin) discovered that if one exposes a current conducting body to a magnetic field the electrical resistance changes. This effect, called the *magnetoresistive effect*, was only used much later for realising transducers. It was not until the American physicist E. F. Hall had discovered the so-called Hall effect that an explanation for this phenomenon could be given. As both effects occur most strongly in semiconductors, they became important for measurement instrumentation only after the development of semiconductor technology. The Hall effect is a result of the Lorentz force, exerted on charge carriers in solids. When a platelet of conducting material is positioned in a magnetic field (as sketched in Fig. 3.19), the charge carriers will be deflected in a direction perpendicular to the direction of motion of these charge carriers and perpendicular to the induction vector **B** of the magnetic field. The Lorentz force acting on a charge q with velocity **v** is equal to:

$$\mathbf{F}_l = q(\mathbf{v} \times \mathbf{B})$$

As the charge carriers are deflected, a transverse charge gradient will build up, resulting in an electrical field **E** across the plate. This field will subject the charge carriers to an opposing force $\mathbf{F}_e$, which is given by:

$$\mathbf{F}_e = q\mathbf{E}$$

At a certain point, equilibrium is reached between the Lorentz force and the force produced by the electrical field, so $\mathbf{F}_e = \mathbf{F}_l$ and therefore:

$$\mathbf{E} = \mathbf{v} \times \mathbf{B}$$

Assuming the charge carriers all have approximately the same velocity **v**, the current density **J** is equal to $nq\mathbf{v}$, with n the concentration of charge carriers. If, in addition, we assume that **B** is perpendicular to **v**, as in Fig. 3.19, then $E = JB/nq$. The factor $1/nq$ is called the Hall coefficient and usually denoted by R_H. With $I = bdJ$ and $V = Eb$ we find:

$$V = \frac{1}{nq} \frac{IB}{d} = R_H \frac{IB}{d}$$

For semiconductor materials, in which the majority charge carriers are holes (p-type semiconductors), q is positive and the output voltage of the Hall plate will have the polarity indicated in Fig. 3.19. If the majority charge carriers are electrons (n-type semiconductors) the polarity will be opposite. The Hall coefficient is large for semiconductors, since the concentration n of

Figure 3.19. Hall effect magnetometer. With the indicated directions of **B** and **I**, the polarity of V corresponds to a plate of p-type semiconductor material.

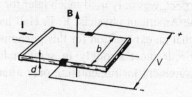

charge carriers is much smaller in semiconductors than in metals.

The assumption that all charge carriers have the same velocity is not entirely correct. Interactions (collisions) with lattice-foreign atoms (impurities) and interactions with lattice defects (dislocations) result in a distribution of the velocity of the charge carriers around a mean velocity. Therefore, the Hall coefficient is found to be somewhere between 0.8 and 1.2 times the theoretical value.

We can explain the Gauss effect or *magnetoresistive effect* with the help of the Hall sensor in Figure 3.19 when we short-circuit the voltage terminals. This creates a current through the sensor between these two terminals. Since the current is perpendicular to the magnetic induction *B*, it will now develop a Hall voltage between the current terminals of the Hall sensor. This voltage is proportional to the forcing current *I* in Figure 3.19 and can therefore be thought as an increase in resistance between the current terminals.

Another way of looking at the consequences of shorting the voltage in a Hall sensor is the following: The restoring electric field **E** which opposes the charge carrier deflection due to the magnetic field inside the sensor (directed from one Hall voltage terminal to the other) is made (nearly) zero by the external short-circuit. This leads to a much stronger deflection of the charge carriers in the sensor. The average path length of the carriers becomes larger, leading to more interactions with the crystal lattice; the resistance as measured between the current terminals of the Hall device becomes greater. Clearly, the magnetoresistance is largest for a short-circuited Hall sensor. There are essentially two ways to realise such a short circuit effectively. One is to use a disc-shaped sensor (Corbino disc) where one terminal is the rim of the disc and the other is the (metallised) centre of the disc (see Figure 3.20). The other is effectively to create an electrical series circuit of a large number of Hall platelets with a large length-to-width ratio. As shown in Figure 3.20, the individual platelets barely develop a Hall voltage.

The Hall effect and the magnetoresistive effect are both *galvano-magnetic effects*. These effects all have in common that they occur in a current carrying conductor when a magnetic field is present. As we have seen above, these galvano-magnetic effects are basically the result of the Lorentz forces acting on the charge carriers in the conductor in the presence of a magnetic field.

Magnetoresistive sensors are often called (magnetic) field plates or Gauss plates. They have the disadvantage of being temperature dependent. The following is an example of the specifications of a Gauss sensor used for field measurement: material InSb, NiSb, $R(0) = 50 \ \Omega \pm 20\%$. The temperature coefficient of $R(B)$ depends on B: when $B = 0$ T it is $-2 \times 10^{-3}\,\mathrm{K}^{-1}$ and when $B = 1$ T it is $-6 \times 10^{-3}\,\mathrm{K}^{-1}$.

Hall plates are frequently used for measuring magnetic fields. The linearity of $V = V(B)$ can be increased further by connecting a resistor R_l to the voltage terminals of the Hall plate, as indicated in Fig. 3.21(a). This linearisation is possible because the internal resistance R_i of the Hall voltage source depends on the induction B, as shown in Fig. 3.21(b). This partially results from the Gauss effect arising from the conducting connector strips along each side of the Hall current terminals of the plate, which short-circuit the sides of the Hall plate.

The voltage terminals of a Hall plate are usually not positioned exactly opposite each other. This will cause a non-zero output voltage, even when the induction B is zero. The magnitude of this zero offset error, the so-called *ohmic zero-offset error*, is proportional to the current I. Therefore, the zero error is usually compensated by a voltage which is derived from the current I. If the current I is zero and the Hall plate is placed in an alternating magnetic field, the output voltage will again usually differ from zero. The alternating field will induce a voltage in the loop which is formed by the

Figure 3.20. The resistance $R(B)$ of differently shaped magnetoresistive sensors as a function of the induction B.

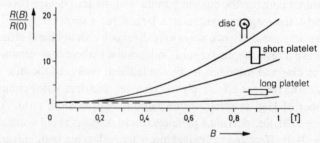

Figure 3.21. (a) Linearisation of the characteristic of $V = V(B)$ by R_l.
(b) Variation of the internal resistance R_i of the Hall voltage source V as a function of the induction B.

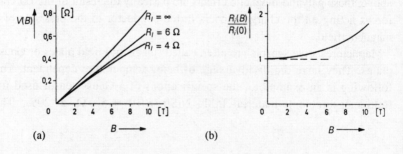

wires connected to the voltage terminals of the Hall plate. This voltage is called an *inductive zero-offset error*. The offset error can be minimised by twisting the wires.

The following is an example of the specifications of a Hall plate: material InAs, sensitivity 60 mV/T at a supply current of 100 mA; temperature coefficient of the sensitivity -10^{-3} K^{-1}; internal resistance $R_i = 1.5$ Ω with $B = 0$ T; load resistance R_l for a linear characteristic of $V(B)$ $R_l = 6$ Ω,;non-linearity less than 1% over the range $0 < B < 1$ T.

Obviously, a Hall plate can measure static magnetic fields, without any moving or rotating parts. A Hall plate is also suitable for high-frequency measurements; it has a wide frequency range (up to several GHz). Furthermore, the disturbance to the magnetic field, caused by the Hall plate is very small. In addition to measuring magnetic fields directly, Hall plates are often used for measuring large DC currents and in current probes for oscilloscopes. For measuring DC currents, a ferromagnetic ring, containing two slots in which Hall plates are positioned, is placed around the conductor (Fig. 3.22). The output voltages of these two plates are summed, so that disturbances by external magnetic fields are eliminated. The Hall voltages resulting from an external induction B are of equal magnitude but of opposite sign, provided that the sensitivities of both Hall plates are equal. This method therefore compensates any Hall voltages caused by external interference fields.

3.3 Signal conditioning

Usually a measurement signal must undergo some form of signal conditioning before it is suitable for displaying, recording or control. This section will deal with several methods of signal conditioning frequently used in measurement instrumentation. We will distinguish between *linear signal*

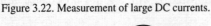

Figure 3.22. Measurement of large DC currents.

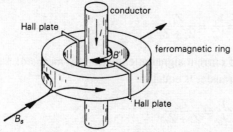

conditioning — such as attenuation, amplification compensation, etc. — and *non-linear signal conditioning* — such as the determination of the peak value, the mean or the RMS value of a signal. In addition, we will also consider *signal conversion* — for instance, sampling, analogue-to-digital conversion and digital-to-analogue conversion. Signal conversion serves to map the signal from a continuous time and amplitude signal into a discrete time and/or discrete amplitude signal or *vice versa*. We will first discuss several types of linear signal conditioning.

3.3.1 Attenuators

If a measurement signal is too large to be measured directly, an attenuator is used. This is to prevent distortion of the signal and the associated loss of measurement information. An attenuator shifts the input dynamic range of a measurement system to higher signal levels. An attentuator can be realised in the form of a resistor network (resistive attenuators), except when dealing with extremely large currents or high voltages, as then the dissipation becomes too large. In the latter case *inductive* or *capacitive attenuators* are used. An additional requirement for this large current, high-voltage type of attenuator is that the measurement instrument must be galvanically insulated from the measurement object (for instance, when performing measurements on high voltage electric power distribution cables). For this reason, transformers are the most common form of inductive attenuators. We will now discuss several types of attenuating network.

Input attenuators

The measurement signal can often most easily be attenuated directly at the input of the measurement instrument by utilising its input impedance Z_i. If the signal is, for instance, a voltage signal it can be attenuated by means of a series impedance Z_s placed in series with the input impedance. If the signal is a current, a parallel, or shunt impedance, Z_p is connected across the input. From Fig. 3.23 the voltage transfer β_v with a series impedance at the input follows as:

$$\beta_v = \frac{V_i}{V_o} = \frac{Z_i}{Z_i + Z_s}$$

In the case of current signal attentuation by means of a shunt impedance Z_p the current transfer is equal to:

$$\beta_i = \frac{I_i}{I_o} = \frac{Z_p}{Z_i + Z_p}$$

The attenuation is by definition equal to the magnitude of the reciprocal of the transfer function. Voltage measurements require a large $|Z_i|$ and therefore R_i of the measurement system has to be large. As a consequence, the (parasitic) input capacitance C_i (in parallel with R_i) can considerably reduce $|Z_i|$, especially at high frequencies. We wish to accomplish an attenuation which does not depend on the input signal's frequency. This can be achieved by shunting the inserted series resistance R_s with a capacitor C_s, so that $R_s C_s = R_i C_i$. Then the attenuation will not depend on the frequency and it becomes:

$$\frac{1}{\beta_v} = \frac{R_i + R_s}{R_i}$$

The use of C_s is referred to as *frequency correction* of the attenuator. It is used, for instance, in the voltage signal attenuator probes for oscilloscopes.

For measuring a current, Z_i must be small. For small values of R_i, frequency dependent attenuation results mostly from the parasitic series induction L_i in series with R_i. This frequency dependence can be eliminated by connecting an inductor L_p in series with R_p, so that $L_p/R_p = L_i/R_i$. The attenuation is now constant for all frequencies and is given by:

$$\frac{1}{\beta_i} = \frac{R_i + R_p}{R_p}$$

Figure 3.23. (a) Voltage signal attenuation by means of a series impedance Z_s. (b) Current signal attenuation with the help of a shunt impedance Z_p.

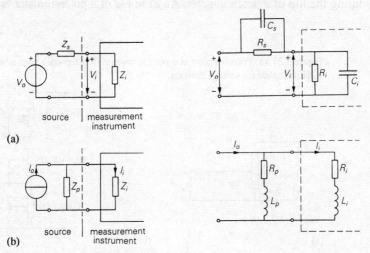

Shunt resistors (with built-in frequency correction) are used to enlarge the measurement range of current measuring instruments.

The method of attenuation which utilises the input impedance of the measurement instrument is a simple and low-cost method. It has the advantage that for voltages the input impedance is increased and for currents the input impedance is decreased, thus reducing loading of the measurement object. A disadvantage, however, is that the input impedance Z_i is not always precisely known, and therefore the measurement becomes less accurate. Moreover, it is difficult to realise a large attenuation factor with this method, since this would require very large series impedances and extremely small shunt impedances. Therefore, separate attenuator networks are used for an accurate attenuation or for a large attenuation.

Voltage dividers

A voltage attenuator which is connected between a (low-impedance) voltage source and a high-impedance instrument input is usually called a *voltage divider*. The terms 'low' and 'high' impedance must be considered relative to the input and output impedances of the voltage divider network. If the requirements for the impedances are not satisfied, an error will occur which can be larger than the inaccuracy of the voltage divider itself.

Fig. 3.24(a) shows a simple *potentiometer* used as a voltage divider. When the potentiometer is not loaded, the voltage transfer β_v is equal to θ. However, if the output is loaded, the transfer β_v will no longer be a linear function of the position of the wiper. A potentiometer is normally not suitable when an accurate adjustable attenuation is needed, since the resistive film or the resistance wire of the potentiometer wears significantly during the life of a potentiometer. An example of a potentiometer is: wire-

Figure 3.24. (a) Potentiometer as a voltage divider. (b) Step-wise adjustable resistor decade for voltage division.

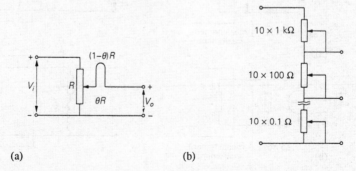

(a)   (b)

wound ten-turn potentiometer; range from 100 Ω to 100 kΩ, non-linearity 10^{-3} (no load); resolution 10^3–10^4.

For accurate, adjustable voltage division a number of resistor decades with step-wise adjustable values are used. If the resistor values increase in factors of 10 the attenuator is referred to as a *decade resistor bank* (see Fig. 3.24(b)). The greatest weakness of such a resistor bank is caused by the switches. The contact resistance must be very low, especially in the lower decades. The thermovoltages of the contacts must also be kept to a minimum. The resolution depends on the number of decades n used and is equal to 10^n. A resistor bank can easily achieve an inaccuracy of 10^{-4}–10^{-5}.

A disadvantage of a resistor bank is the fact that the input impedance is not constant and, for this reason, other types of voltage dividers are often preferred. An example of one of these is given in Fig. 3.25: the Kelvin–Varley voltage divider. The last decade of the illustrated divider has ten steps of 0.8 Ω. The total resistance of 8 Ω of this decade is connected in parallel to two resistors of 4 Ω of the previous decade. The one but last decade consists of eleven resistors of 4 Ω. Its output is always tapped across two of these resistors. Therefore, the number of steps of this decade is also ten. The input resistance of this stage is therefore not 11×4 Ω but 10×4 Ω, due to the parallel connection of the last stage. The input resistance of the second decade is 200 Ω and that of the first decade is 1 kΩ. Thus, the input resistance is independent of the set voltage division. Another advantage of the Kelvin–Varley voltage divider is that the current in each succeeding decade is halved, which means that a relatively high resistance level can be used, even for the last decades, reducing the influence of the switch contact resistance.

Characteristic attenuators
In high-frequency or wide-band measurement systems which are designed to operate with a certain characteristic impedance (for eliminating reflections

Figure 3.25. A Kelvin–Varley voltage divider.

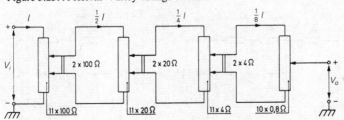

or standing waves on transmission lines), *characteristic attenuators* must be used. This type of attenuator only provides the desired attenuation when it is used in conjunction with a specified (characteristic) impedance. The attenuator must therefore be driven by a signal source with a particular output impedance and must also be loaded with a particular (often identical) impedance. A simple realisation of this is a characteristic attenuator consisting of a number of cascaded T-sections. Fig. 3.26 shows a T-section of such an attenuator. Normally in these cases, $R_v = R_c$ and $R_l = R_c$, where R_c is the characteristic resistance. R_v is the source resistance and R_l the load resistance. Therefore, the input resistance R_i of the T-section must be equal to R_c when a load is connected and the output resistance R_o must be equal to R_c when a source with $R_s = R_c$ is connected to the T-section. Since $R_i = R_c$ and $R_o = R_c$, it follows that $R_1 = R_2 = R_s$ and $R_3 = R_p$ (see Fig. 3.26(b)). The relation for R_s and R_p can now be found:

$$R_i = R_c = R_s + \frac{(R_s + R_c)R_p}{R_s + R_c + R_p}$$

so $R_c^2 = R_s^2 + 2R_sR_p$.

The voltage transfer β_v follows from:

$$\beta_v = \frac{V_o}{V_i} = \frac{R'}{R_s + R'} \frac{R_c}{R_s + R_c}$$

With $R' = R_p//(R_s + R_c)$ and $R_p = (R_c^2 - R_s^2)/2R_s$ this results in:

$$\beta_v = \frac{R_c - R_s}{R_c + R_s}$$

The attenuation of a characteristic attenuator is most often expressed in dB (see Appendix A.3). We can write the attenuation α in dB:

Figure 3.26. (a) Characteristic attenuator consisting of a single T-section. (b) Symmetrical T-section. R_c is the characteristic resistance.

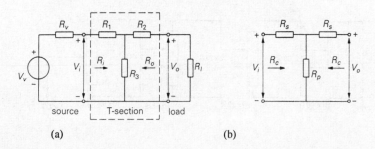

(a)   (b)

$$\alpha = 10 \log_{10}\left(\frac{P_1}{P_2}\right) = 20 \log_{10}\left(\frac{V_i}{V_0}\right)$$

since $R_v = R_i = R_c = R_o = R_l$. Thus, we find:

$$\alpha = -20 \log_{10}(\beta_v) \text{ [dB]}$$

The advantage of characteristic attenuators is that they can be immediately cascaded. The total voltage transfer β_v is equal to the *product* of the transfer factors of the separate sections. If the attenuation is expressed in dB, the total attenuation is equal to the *sum* of the attenuation of the separate sections. So:

$$\beta_{tv} = \prod_{i=1}^{n} \beta_{iv} \quad \text{and} \quad \alpha_t = \sum_{i=1}^{n} \alpha_i$$

in which β_{tv} is the total voltage transfer and α_t is the attenuation (in dB) of a compound characteristic attenuator, consisting of n characteristic sections.

An example of a characteristic attenuator with three cascaded symmetrical T-sections is: $n = 3$, $R_c = 60 \ \Omega$, $\beta_{tv} = \frac{1}{8}$ or $\alpha_t = 18$ dB. Thus, for one section $\beta_v = \frac{1}{2}$ and $\alpha = 3.6$ dB, so $R_s = 20 \ \Omega$ and $R_p = 80 \ \Omega$. It is also possible to assemble characteristic attenuators from sections other than T-sections, such as π- or H-sections (see Fig. 3.27).

Measurement transformers

For attenuating high voltages and large currents transformers are usually used since the dissipation of resistive attenuators would become excessive. For reasons of safety, transformer attenuators are usually preferred to inductive or capacitive attenuators because of the insulation a transformer provides between the measurement object and the instrument. In instrumentation transformers are frequently used for purposes other than power measurement applications. For instance, driving 'floating' bridge networks, noise matching of low impedance sources and avoiding ground loops is

Figure 3.27. Characteristic attenuators consisting of π- and H-sections.

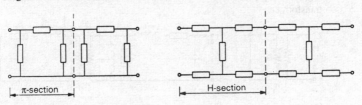

π-section   H-section

often done with a transformer. In addition, they can also be found in current probes for oscilloscopes, in multimeters and in many other measurement instruments. In the following discussion, we will limit ourselves to measurement transformers for power applications.

Let us first consider an ideal transformer (see Fig. 3.28(a)). It is defined by:

$$\frac{V_2}{V_1} = \frac{n_2}{n_1} = n$$

$$\frac{I_2}{I_1} = -\frac{n_1}{n_2} = -\frac{1}{n}$$

For an ideal transformer, the *turns ratio n* entirely determines both the voltage transfer and the current transfer. The voltages are in phase and the currents are 180° out of phase. In reality, though, this is only partially true . The deviation from ideal behaviour is characterised by the transfer magnitude error ε and the phase error θ.

The *transfer error ε_v* for the voltage is defined as:

$$\varepsilon_v = \frac{V_2/n - V_1}{V_1}$$

and the transfer error ε_i for the current as:

$$\varepsilon_i = \frac{nI_2 + I_1}{I_1}$$

since $I_1 \approx -nI_2$.

The *phase error θ_v* is equal to the phase difference between the input and output voltages. The phase error θ_i of the current transfer function is the deviation of the current phase angle difference from 180°. The polarities of the voltages and the directions of the currents are defined as in Fig. 3.28(a). The transfer and phase errors are caused by various parasitic effects. Therefore, we must add a number of extra impedances to the ideal transformer model to obtain the electrical analogue of a practical measurement

Figure 3.28. (a) Ideal transformer. (b) A more realistic electrical analogue of a transformer.

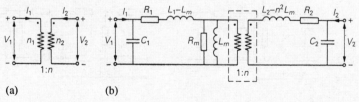

(a)   (b)

transformer (see Fig. 3.28(b)).

In this analogue, L_1 is the inductance of the primary and L_2 that of the secondary. The magnetisation current which generates the flux in the core flows through the inductor L_m, which is equal to $L_m = M/n$ with $M = k\sqrt{L_1 L_2}$ The coupling coefficient k determines the fraction of the flux generated by the primary that is enclosed by the secondary. If $k = 1$, the inductors $L_1 - L_m$ and $L_2 - n^2 L_m$ will both become zero. These inductors represent the flux lost due to leakage (leakage inductances). The resistors R_1 and R_2 represent the resistance of the primary and the secondary (copper losses) and the resistor R_m characterises the losses in the core of the transformer (iron or core losses). Finally, the capacitors C_1 and C_2 form the total capacitance of the turns. Usually C_1 and C_2 may be neglected. However, at high frequencies, C_2 may resonate with $L_2 - n^2 L_m$, resulting in a resonant peak in the frequency transfer function of the transformer. This high-frequency peak is sometimes referred to as the leakage resonant peak (due to the leakage inductance).

The values of the impedances in Fig. 3.28(b) are determined by the design and construction of the transformer and the nature of the ferromagnetic core material. The design of a transformer can be optimised for a minimal current transfer error and a minimal current phase error for a current transformer, or, it can be optimised for minimal voltage transfer and voltage phase error for a voltage transformer. In the secondary of a *voltage transformer* errors occur when the transformer is loaded. They are caused by the presence of R_2 and $L_2 - n^2 L_m$. Errors in the primary are caused by this load current being transformed into the input and by the magnetisation current (due to L_m and R_m). Therefore, it is important to keep the load current of a voltage transformer as small as possible. The magnetisation current is minimised by using magnetic core material with a high μ_r and no air gaps. R_m is kept large by keeping the maximal allowable induction small. The transformer is wound so as to allow only a minimum leakage, for instance, by winding on top of the secondary over the primary onto a toroidal core. This will reduce both leakage inductances, $L_1 - L_m$ and $L_2 - n^2 L_m$.

A voltage transformer may never be shorted; this would lead to destructive current densities and overheating.

A *current transformer* has only a few (for large currents only one) turns for the primary. Since the current transformer is nearly short-circuited, the voltage of the primary or the secondary is never more than a few volts. The transformer is fed from a current source (current driven). The impedances R_1 and $L_1 - L_m$ have little effect on the transfer. However, the secondary resistance and leakage inductance must remain small. This is achieved with thick wire and a good coupling between the primary and secondary. If the secondary were left open-circuited, the voltage across L_m would increase sharply, and with it the degree of saturation of the magnetic circuit. The losses due to hysteresis would increase, the transformer would become very hot and would finally be destroyed. The cause of failure can also be break-down of the insulation between the windings due to the sudden increase of the primary voltage. Therefore, a current transformer must always be (nearly) short-circuited during operation. For current and voltage measure-ments with transformers only the transfer errors ε are important. However, for power measurements not only the transfer error ε, but also the phase error θ must be taken into account, since the power is equal to $P = VI \cos \phi$, where ϕ is the angle between V and I.

Transformers are classified according to their inaccuracy. The 0.1 class has a transfer error smaller than $\pm\,0.1\%$ and a phase error smaller than $\pm\,5$ minutes for a load impedance with a $\cos \varphi$ between 1 and 0.8. Class 1 has a transfer error less than $\pm\,1\%$ and a phase error smaller than 60 minutes under the same conditions.

Voltage transformers are available for voltages from 1 kV to 500 kV. The output voltage is standardised to a nominal value of 100 V and $100\sqrt{3}$ V. Current transformers are available for 1 A to 50000 A. The output current is standardised to 1 A and 5 A.

3.3.2 Compensator networks

A compensator is a measurement network which makes use of the compen-sation method. Its major advantage is that it can make measurements *without loading* the measurement object with a high *accuracy*. Nowadays compensator networks are no longer used very frequently. Thanks to modern electronics we now have digital instruments for measuring voltage and current which allow measurements to be performed much more easily and extremely accurately. Some of these instruments make use of the *compensation method* internally too. The compensation method does still have a number of other applications (such as in pen recorders and other servo systems). These systems can be designed such that they compensate the measured quantity without the intervention of a human operator: *automatic compensation*.

First, though, we will consider measurement systems in which the unknown quantity is compensated by a human observer: *non-automatic compensation*, as shown in Fig. 3.29. Obviously, one can compensate both across quantities and through quantities. In Fig. 3.29(a) the zero indicator

ΔV is used to adjust the auxiliary voltage source V_a so that $\Delta V = 0$. In this zero state the through quantity becomes zero, so there cannot be any exchange of energy between V_i and V_a. (Of course, here we are using the symbol of the generalised across quantity V.) The set value of V_a is equal to the value of V_i. As this set value is known, the measured quantity is also known. Fig. 3.29(b) shows a flow diagram for this process. The auxiliary quantity V_a is subtracted from the input quantity V_i and the difference from the desired situation, $\Delta V = 0$, is read out. This is used to readjust V_a until $\Delta V \approx 0$. The output quantity (the measurement result) is the set value of V_a when the zero state is reached.

Fig. 3.29(c) demonstrates how a through quantity I can be compensated. The zero state is reached here when $\Delta I = 0$. The across quantity at the terminals of the null indicator is then zero and, consequently, the current through R_i and R_a is zero. Therefore, in this zero situation $I_i = I_a$. Again, the output quantity is equal to the setting of I_a. Comparing Figs. 3.29(b) and 3.29(d) we see that the signal flow diagrams are identical.

Automatic compensation, i.e. compensation without human intervention, is shown in Fig. 3.30. The symbols V_i, ΔV, V_a and I_i, ΔI, I_a correspond to those in Fig. 3.29. A voltage amplifier is given as an example of the compensation of an across quantity. Assuming that the gain A of the amplifier is extremely large, the zero state will automatically adjust to $\Delta V \approx 0$. Then $V_i = V_a$. Since $V_a = V_o R_1/(R_1 + R_2)$, the voltage amplification is equal to $V_o/V_i = (R_1 + R_2)/R_1$. Fig. 3.30(b) depicts a block diagram for this automatic compensation.

Fig. 3.30(c) shows a current-to-voltage amplifier (often, incorrectly, called a current amplifier) as an example of automatic compensation of a through

Figure 3.29. Non-automatic compensation. (a) Compensation of an across quantity V with (b) the corresponding flow diagram. (c) Compensation of a through quantity I with (d) the associated flow diagram.

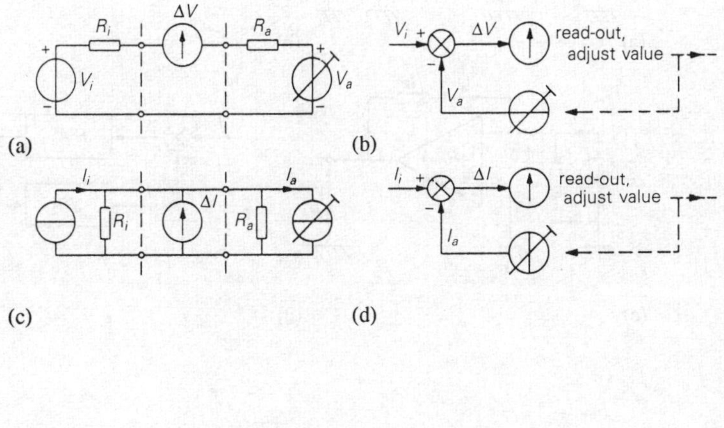

quantity. Assuming the input resistance R_i and the amplification A of the amplifier are extremely large, even a small difference in current ΔI would cause a large voltage at the input of the amplifier, which, after amplification, would appear as an exceedingly large voltage at the output. Since V_o is limited by the supply voltage, the input voltage and the difference current ΔI will be very small. Thus, $I_a \approx I_i$. The auxiliary current is generated from the output voltage by means of R_2, so $I_a = V_o/R_2$. With the polarity of V_o and the direction of I_i as indicated in Fig. 3.30(c), the transfer function is equal to R_2. The corresponding block diagram is given in Fig. 3.30(d). This is similar to Fig. 3.30(b). In both cases, compensation of the input quantity is accomplished by applying *negative feedback*. The associated feedback loop can also be found in Figs. 3.29(b) and 3.29(d), though there, the human observer is part of the forward path. In the case of automatic compensation, the human observer/controller has been replaced by a system with a very high loop gain. We can consider *feedback* as a form of *continuous automatic compensation*. The compensation here does not occur once or at isolated points in time, but is achieved continuously.

3.3.3. Measurement bridges

In the discussion of measurement methods in Section 2.2, we saw that the

Figure 3.30. Automatic compensation. (a) Compensation of an across quantity V_i with (b) the corresponding block diagram. (c) Compensation of a through quantity I_i with (d) the corresponding block diagram.

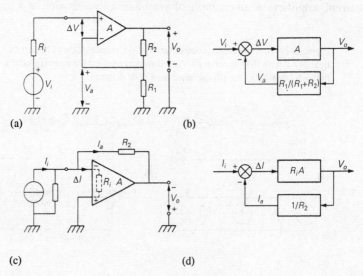

bridge method is often used for measuring or comparing impedances. These impedances may also be generalised or non-electrical. An example of a mechanical bridge network is given in Fig. 3.31. The masses m and m' are suspended from two damped springs. The bar, to which the springs are attached, is given a velocity V and the difference in velocity V_d between the masses is measured with an inductive velocity sensor (see Section 3.2.1). The mass of this sensor may be neglected. The analogue shown in Fig. 3.31(b) shows that this method can easily compare the mass-spring systems and allows us to adjust them such that they are equal (for example, suspension tuning in a vehicle).

A second example is a thermal bridge network, shown in Fig. 3.32. The dotted materials exhibit a thermal impedance equal to $Z = \Delta T / I_h$, where I_h is the heat flow through the material. Unfortunately, the heat flow is rather difficult to measure, so it is hard to determine Z from the above definition.

Figure 3.31. (a) Mechanical bridge network. The velocity difference between mass m and mass m' is converted into an electrical signal; the output signal of the bridge. (b) Analogue of the network of (a). V is the bridge input velocity.

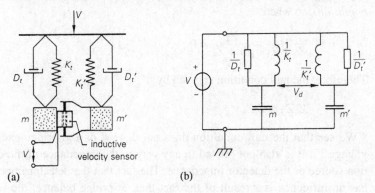

(a)   (b)

Figure 3.32. A thermal bridge network. T_0 is the ambient temperature, ΔT is the bridge excitation temperature.

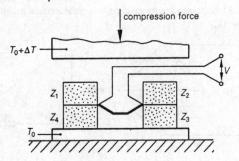

Therefore, the unknown thermal impedance Z_1 in Fig. 3.32 is compared to three known thermal impedances Z_2, Z_3 and Z_4. When the difference in temperature (as measured with a thermocouple) is zero for a sinusoidal bridge excitation ΔT, Z_1 can be expressed as a function of Z_2, Z_3 and Z_4 at that frequency.

Measurement of impedances, in general, is based on a bridge network with three known impedances and one unknown impedance. One or more of the known impedances is adjustable. Fig. 3.33 illustrates that a bridge network can be considered as a two-port network. The input voltage V_i is equal to the bridge supply voltage V_s and the output voltage V_o is equal to the detector voltage V_d. If the internal impedance of the bridge supply voltage source is negligible with respect to $(Z_1 + Z_4)//(Z_2 + Z_3)$ and the impedance of the null detector is much larger than $(Z_1//Z_4) + (Z_2//Z_3)$, we can find an expression for the transfer of the network. The detector voltage is given by:

$$V_d = V_s \left(\frac{Z_3}{Z_2 + Z_3} - \frac{Z_4}{Z_1 + Z_4} \right)$$

The bridge is balanced when the transfer is equal to zero (*balance or null condition*), so when:

$$\frac{Z_3}{Z_2 + Z_3} - \frac{Z_4}{Z_1 + Z_4} = 0$$

Therefore, the null condition is given by:

$$Z_1 Z_3 = Z_2 Z_4$$

We see that the null condition does not depend on the bridge excitation voltage V_s. It is also not related in any way to the impedance of the excitation source or the detector impedance. The fact that the detector impedance has no influence is a result of the fact that, at bridge balance, the voltage across the detector is zero. The fact that the null condition is independent of

Figure 3.33. Schematic representation of a measurement bridge network as a two-port network.

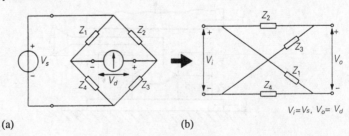

$V_i = V_s, \ V_o = V_d$

(a)   (b)

the source impedance can be explained by arguing that the source impedance will only decrease the voltage V_i of the bridge network, as the supplied voltage will be divided between the source impedance and the input impedance of the bridge network.

An important parameter of an impedance bridge is the *sensitivity*. The bridge sensitivity determines the magnitude of the null detector voltage or current for a small deviation from the balanced condition. Assuming the bridge is balanced by means of the adjustable impedance Z_1 in Fig. 3.33, the differential sensitivity S_{diff} of the detector voltage V_d can then be written as:

$$S_{\text{diff}} = \frac{dV_d}{dZ_1} = V_s \frac{Z_4}{(Z_1 + Z_4)^2}$$

Since the absolute magnitude of Z_1 can differ from bridge to bridge, we are more interested in the sensitivity of V_d to a relative change dZ_1/Z_1 in this impedance Z_1. This yields:

$$S_B = \frac{dV_d}{dZ_1/Z_1} = \frac{dV_d}{dZ_1} Z_1 = V_s \frac{Z_1 Z_4}{(Z_1 + Z_4)^2}$$

If we now introduce the *bridge ratio* F ($= Z_2/Z_3$), and, of course, when the bridge is balanced $F = Z_1/Z_4$, the sensitivity S_B can be written as:

$$S_B = V_s \frac{F}{(1 + F)^2}$$

For impedance bridges the bridge ratio F tends to be a complex quantity. In

Figure 3.34. The magnitude of the bridge sensitivity $|S_B|$ as a function of the complex bridge ratio F.

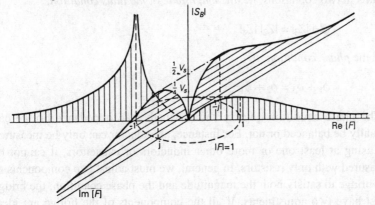

Fig. 3.34 the bridge sensitivity S_B is plotted as a function of F. If F is positive and real, $|S_B|$ is at its maximum when $F = 1$. The sensitivity is then $|S_B| = V_s/4$. The sensitivity of a Wheatstone bridge, for instance, is maximal when the resistance of the pair of resistors above the null detector and that of the pair of resistors below the null detector are the same (pair wise). The bridge sensitivity has a pole at $F = -1$.

It is often possible to increase the sensitivity of a bridge by exchanging the detector and the bridge excitation voltage source. Generally speaking, the detector must be connected between the node in which the two largest impedances meet and the node in which the two smallest impedances meet. The excitation voltage source is then connected between the two remaining nodes.

It is evident that the voltage sensitivity of a bridge network will decrease as the internal resistance of the null detector decreases: the loading of the bridge will become larger.

It is also obvious that one can increase the sensitivity of a bridge by increasing the supply voltage V_s. V_s is only limited by the maximum allowable dissipation in the bridge components (and so by the maximum component temperature that does not (yet) affect the impedance of the components too much).

When the bridge impedances are complex, i.e. when:

$$Z_i = |Z_i|\, e^{j\phi_i} \quad (i = 1, 2, 3, 4)$$

the balance condition:

$$Z_1 Z_3 = Z_2 Z_4$$

results in two conditions: i.e. the *magnitude* or *modulus condition*:

$$|Z_1|\,|Z_3| = |Z_2|\,|Z_4|$$

and the *phase condition*:

$$\phi_1 + \phi_3 = \phi_2 + \phi_4$$

The phase condition quickly shows whether an impedance bridge can actually be balanced or not. For instance, an inductor can only be measured by using at least one or more other inductors or capacitors, it cannot be measured with only resistors. In general, we must adjust two components of the bridge to satisfy both the magnitude and the phase condition: the bridge must have two adjustments. If all the components of the bridge are ideal resistors, capacitors or inductors, the complex balance condition is reduced to a magnitude condition only. For example, in a Wheatstone bridge measuring resistance, only the magnitude condition is left: $R_1R_3 = R_2R_4$. Only one of these components needs to be adjustable.

The accuracy of a bridge measurement is determined almost solely by the accuracy of the bridge impedances, as the unknown impedance is expressed in terms of the other three impedances. The variable impedances are often adjustable only in steps for better accuracy.

The resolution of the adjustment of the bridge components must be sufficiently large to achieve the required accuracy. In addition, the null detector must be sensitive enough to be able to detect a variation in a bridge component as small as the smallest adjustment step.

The accuracy may also be affected detrimentally by parasitic impedances associated with the bridge network itself and with the components in the network. *Network parasitic impedances* result from parasitic coupling both mutually between bridge components and between bridge components and a ground connection. These impedances can be fixed and subsequently eliminated by grounding and screening the bridge correctly. Component parasitic impedances are due to imperfections of the bridge components, such as series resistance, lead inductance, etc. The magnitude of these impedances is mainly determined by the construction of the bridge components.

An example of a method avoiding parasitic network impedances is provided in Fig. 3.35, where a so-called *Wagner ground* is depicted. The bridge network, which consists of the bridge components Z_1, Z_2, Z_3 and Z_4, exhibits stray capacitance to ground, denoted by the lead capacitances C_A, C_B, C_C and C_D. These stray capacitances will affect the balance of the bridge. This effect can be eliminated by grounding the entire bridge network at the centre tap of two extra impedances Z_5 and Z_6. First, we balance the

Figure 3.35. Wagner grounding.

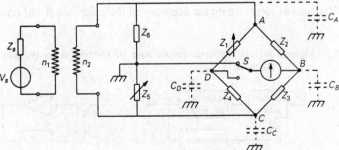

bridge, consisting of Z_1, Z_2, Z_3 and Z_4, with Z_1, after which we switch S and again balance the bridge, which now consists of Z_2, Z_3, Z_5 and Z_6, with, for instance, Z_5. This process is then repeated several times until both bridges are perfectly balanced. When this is achieved, the voltage across the capacitors C_B and C_D will be zero with respect to ground. The capacitors C_A and C_C are effectively circuited in parallel to Z_6 and Z_5 and will therefore play no role in the balance condition of the actual bridge, Z_1, Z_2, Z_3 and Z_4.

3.3.4 Instrumentation amplifiers

When a measurement signal is too small, it is usually first amplified. Accurate, low-noise and low-distortion amplification is performed by *instrumentation amplifiers*. This special type of amplifier usually has an accurately defined, adjustable gain. Besides increasing the sensitivity of the measurement, the instrumentation amplifier also isolates the measured object from the load of the measurement system. An instrumentation amplifier will increase the power of the measurement signal, whilst preserving the information content. The energy extracted from the measurement object can therefore be considerably reduced. For measuring a voltage, the input impedance of the instrumentation amplifier must be high and for measuring a current it must be low, to prevent loading of the measurement object.

As we have already seen in Section 3.3.2 it is possible to compare a physical quantity with another quantity having the same physical dimension, without loading the measurement object by means of compensation networks. In Fig. 3.36 it is shown once more that this can be accomplished automatically and continuously in time. The measurand is the input quantity V_i, the other quantity is the set value or output quantity V_o. This is adjusted to produce a minimal error $V_e = V_i - V_r$. Then $V_r \approx V_i$. In general V_r will not be equal to V_o; the relation between V_o and V_r will be determined by the transfer characteristics of the reference β. The accuracy of V_o as a measure for V_i depends therefore on the accuracy of the reference β and on how well

Figure 3.36. (a) Continuous compensation. (b) Continuous compensation using negative feedback.

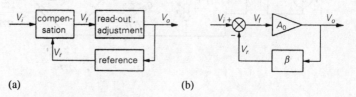

(a)   (b)

the input quantity is compensated by V_r. Compensation will never be complete. With automatic compensation errors are caused by the finite value of $A_0\beta$; there must always be a non-zero error signal V_e for the amplifier gain A_0 to be able to produce the output signal V_o.

A high accuracy is achieved by a large gain A_0 in the forward path and an accurately determined feedback factor β. This is also evident from the following expression. The transfer function of the negative feedback amplifier of Fig. 3.36(b) is given by:

$$A_t = \frac{V_o}{V_i} = \frac{A_0}{1 + A_0\beta}$$

For *negative feedback* we must always ensure that $|1 + A_0\beta| > 1$. When $|1 + A_0\beta| < 1$ the feedback loop will give positive feedback. When $|1 + A_0\beta| = 1$ there is no feedback. With a strong negative feedback, i.e. when $|1 + A_0\beta| \gg 1$, the overall gain of the system will approximate $A_t \approx 1/\beta$. Therefore, for the signal actually to be amplified, β must be smaller than 1 (attenuation in the backward path). For a strong feedback $|A_0\beta| \gg 1$. Evidently, the feedback loop reduces a large gain A_0 to a much smaller gain A_t which, however, is almost entirely determined by the feedback factor β; $A_t \approx 1/\beta$. The overall gain A_t will always differ slightly from the desired value $1/\beta$. If ε is the relative difference between A_t and $1/\beta$, i.e. if:

$$A_t = \frac{1}{\beta}(1 - \varepsilon)$$

this relative difference is determined by:

$$\varepsilon = \frac{1}{1 + A_0\beta}$$

Thus, for negative feedback $|\varepsilon| < 1$ and for positive feedback $|\varepsilon| > 1$. In the case of negative feedback, the relative error $|\varepsilon|$ will become smaller as the so-called loop gain $A_0\beta$ is increased. If, for some reason or other, for instance due to a multiplicative disturbance, A_0 changes, it will produce only a minor change in A_t. In Section 2.3.3.3 we calculated that:

$$\frac{\Delta A_t}{A_t} = \varepsilon \frac{\Delta A_0}{A_0}$$

Therefore, for measurement purposes, we use strong negative feedback, determined by an accurate, time and temperature stable attenuator, insensitive to disturbances. It must be noted that applying feedback to an open-loop amplifier will not reduce the *additive distortion* in the amplifier. Also, the application of feedback will affect the input and output impedances of the amplifier. This is illustrated by the instrumentation amplifier shown in Fig. 3.37, which is used for measuring small electrical currents. The input current is here converted into a proportional output voltage. A simple calculation shows that the transfer function of this circuit is given by:

$$\frac{v_o}{i_i} = -\frac{R_i'(A_0 R - R_o')}{R_i'(1 + A_0) + R + R_o'}$$

An amplifier without the feedback loop is usually realised as a so-called *operational amplifier*. This is an amplifier with an extremely high gain A_0, a very high input impedance R_i' and a low output impedance R_o'. We may therefore assume that $R_i' \gg R$, and that therefore the input current i_i will flow almost entirely through R. Then $i_i = (v_i - v_o)/R$. We may also assume that $R \gg R_o'$, and therefore $v_o \approx -A_0 v_i$, resulting in $i_i = -v_o(1 + A_0)/RA_0$. Obviously, $A_0 \gg 1$, so the transfer function is given by:

$$\frac{v_o}{i_i} = -R$$

provided that $R_i' \gg R \succ R_o'$ and $A_0 \gg 1$. Clearly, this circuit converts an input current i_i into an output voltage v_o. The transfer function has the physical dimension of impedance; this type of amplifier is therefore often referred to as a transimpedance amplifier. The input impedance of this amplifier is very small. We can calculate that the input resistance is given by:

Figure 3.37. Measurement of small electrical currents with a transimpedance instrumentation amplifier.

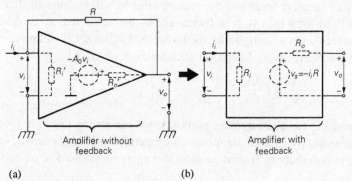

(a)   (b)

$$R_i = \frac{v_i}{i_i} = \frac{R_i'(R + R_o')}{(1 + A_0)R_i' + R + R_o'}$$

The input resistance $R_i = v_i/i_i$ of a current-to-voltage amplifier or trans-impedance amplifier can be calculated with $i_i = (v_i - v_o)/R$ and $v_o = -A_0 v_i$ and is given by:

$$R_i = \frac{R}{1 + A_0}$$

provided that $A_0 R_i' \gg R \gg R_o'$. It turns out that the feedback loop has reduced the input resistance by a factor of $1 + A_0$. The output resistance of the circuit has also decreased.

The expression for the output resistance is:

$$R_o = \frac{v_o}{i_s} = \frac{R_o'(R + R_i')}{(1 + A_0)R_i' + R + R_o'}$$

The current i_s is the short-circuit current which flows out of the positive output terminal of the amplifier. Obviously, in this case, the feedback loop reduces both the input resistance and the output resistance of the circuit by approximately the same factor $1 + A_0$. This can be made clear by letting $A_0 \to 0$. This is equivalent to removing the operational amplifier, but maintaining R_i' and R_o'.

The output resistance of the feedback amplifier can also be determined with Thévenin's theorem: $R_o = v_o/i_s$, with the short-circuit current $i_s = i_i - A_0 v_i/R_o'$. Remembering that $R \gg R_o'$ and $A_0 > 1$, we find $i_s \approx -A_0 v_i/R_o'$. When the output is short-circuited, the input voltage will be $v_i = i_i R R_i'/(R + R_i') \approx i_i R$. So $i_s \approx -i_i A_0 R/R_o'$. The open output voltage is $v_o = -i_i R A_0/(1 + A_0)$, resulting in an output resistance of:

$$R_o = \frac{R_o'}{1 + A_0}$$

provided that $R_i' \gg R \gg R_o'$ and $A_0 > 1$. We see that the operational amplifier with external feedback in Fig. 3.37(a) corresponds to the internally feedback amplifier of Fig. 3.37(b), if $R_i = R/A_0$, $R_o = R_o'/A_0$ and $v_o = -i_i R$.

The following example gives an idea of the order of magnitude of typical values for such a closed-loop amplifier. An operational amplifier with $R_i' = 1\ M\Omega$, $R_o' = 100\ \Omega$ and $A_0 = 10^5$ is used with a feedback resistor $R = 10\ k\Omega$ to construct a current-to-voltage amplifier with an input resistance of $R_i =$

$100\ m\Omega$, an output resistance of $R_o = 1\ m\Omega$ and a transfer function $v_o/i_i = -10\ k\Omega$.

When a voltage signal is too small to be measured directly it can be first amplified by means of a voltage amplifier. This can be performed simply by the circuit of Fig. 3.38(a). This circuit can be derived from the circuit of Fig. 3.37(a) by realising that the circuit in Fig. 3.38(a) is the same as that of Fig. 3.37(a) with a resistor R_1 in series with the input. The input voltage v_i is converted into a current $i_i = v_i/R_1$ with the aid of this resistor R_1, and this current is then converted into the output voltage v_o, as previously. Thus, the total voltage gain is:

$$A_t = \frac{v_o}{v_i} = -\frac{R_2}{R_1}$$

From Fig. 3.37 it follows that the input resistance R_i of this circuit is equal to R_1. This may be inadequate for certain applications, since a voltage amplifier must have an extremely high input resistance. An alternative circuit is therefore given in Fig. 3.38(b). This circuit can also be derived from the previous circuit, by calculating the current through R_1. As the gain A_0 of the operational amplifier is extremely high, the input voltage will be negligible, provided that the amplifier is operating in its linear region with its output voltage somewhere between the amplifier supply voltages. Therefore, the voltage of the common node of R_1 and R_2 will follow (compensate) the input voltage v_i. Then, the current through R_1 must be equal to $i = -v_i/R_1$. This current must be supplied via R_2. The voltage gain of this amplifier therefore becomes:

$$A_t = \frac{v_o}{v_i} = \frac{R_1 + R_2}{R_1}$$

Figure 3.38. Voltage amplifier. (a) Inverting voltage amplifier. (b) Non-inverting voltage amplifier.

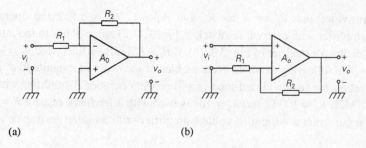

(a)   (b)

The input resistance of this voltage-to-voltage amplifier is far greater than that of the previous amplifier due to the fact that the input voltage is continuously automatically compensated (see also Fig. 3.30(a)). This is not the case in the first circuit of Fig. 3.38(a).

Calculation of the input resistance R_i of the non-inverting voltage amplifier yields:

$$R_i = R_i' \left(1 + \frac{R_1 A_0}{R_1 + R_2 + R_o'} \right)$$

When $R_o' \ll R_1 + R_2$ and $A_0 \gg 1 + R_2/R_1$ this reduces to:

$$R_i \approx R_i' \frac{A_0 R_1}{R_1 + R_2}$$

An operational amplifier for which $R_i' = 1$ MΩ and $A_0 = 10^5$, with a feedback loop designed so that the overall gain $A_t = 100$, has an input resistance of $R_i = 1$ GΩ. The smaller the value for A_t, the larger R_i becomes.

A very strong feedback can be achieved by short-circuiting resistor R_2 and removing R_1 altogether from the voltage amplifier of Fig. 3.38(b). Then, the voltage gain is $A_t = 1$. This unity voltage amplifier is often employed for matching a high output impedance measurement object to a low impedance measurement system. It is extremely well suited for this application as it has an exceptionally high input impedance and very low output impedance. Although the voltage gain of the amplifier is only equal to unity, it can still increase the energy content of a measurement signal tremendously. The power P_i flowing into the unity gain amplifier at the input is $P_i = v_i^2/R_i$. If the input resistance of the measurement system is R_m, this will absorb $P_o = v_v^2/R_m$. Since $v_o \approx v_i$, the power gain amounts to $P_o/P_i = R_i/R_m$. For this reason the circuit in Fig. 3.38 with $R_2 = 0$ and $R_1 = \infty$ is often referred to as a (voltage) buffer circuit.

In Section 2.3.3.3 we saw that disturbances due to ground-loop currents can be avoided by employing a voltage amplifier with a differential input stage. This kind of amplifier is only sensitive to the voltage difference between its input terminals and not to a common-mode signal. Fig. 3.39 shows another application of a differential input amplifier. In this application the strain gauge bridge reading is not affected by the voltage v_1, nor by the voltage v_2, but only by difference voltage v_d. The Thévenin equivalent circuit is given in Fig. 3.39(b). The *difference voltage* v_d is determined by:

$$v_d = v_1 - v_2 = v_s \frac{\Delta R}{R}$$

The *common-mode voltage* v_c is given by:

$$v_c = \frac{1}{2}(v_1 + v_2) = \frac{1}{2}v_s$$

The output resistance R_o of the bridge is equal to $R_o = (R^2 - \Delta R^2)/R$. When $\Delta R \ll R$, this gives $R_o \approx R$. The voltage amplifier connected to the bridge output must only be sensitive to v_d. This type of amplifier is referred to as a *differential amplifier*. The common-mode signal v_c is almost entirely rejected by the amplifier.

A small fraction of the common-mode signal will still reach the output. The gain A_c for the common-mode signal v_c of the differential amplifier is defined as:

$$A_c = \left(\frac{v_o}{v_c}\right)_{v_d = 0}$$

The gain A_d for the differential-mode signal v_d is:

$$A_d = \left(\frac{v_o}{v_d}\right)_{v_c = 0}$$

These two quantities are referred to as the common-mode gain and the differential-mode gain. A measure for the suppression of the common-mode signal is given by the *Common-Mode Rejection Ratio* (CMRR). This ratio is

Figure 3.39. Read-out of a bridge network with a differential amplifier. (a) Bridge network with four active strain gauges. (b) Equivalent circuit; $v_c = v_s/2$, $v_d = v_s$ $\Delta R/R$ and $R_0 = (R^2 - \Delta R^2)/R$. (c) Differential Voltage Amplifier (DVA).

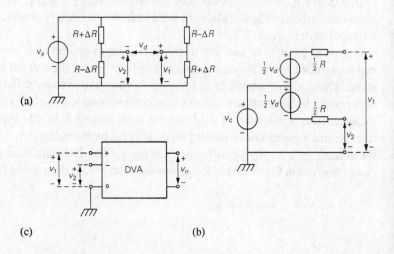

(a)

(c)   (b)

defined as the ratio of the common-mode signal v_c and the differential signal v_d, which would produce the same output signal v_0. In shorthand:

$$\text{CMRR} = \left(\frac{v_c}{v_d}\right)_{\text{same } v_o} = \frac{A_d}{A_c}$$

Depending on the magnitude of the differential amplification A_d, which is often adjustable from 1 to about 10^3, rejection ratios of 10^5–10^8 can be obtained in an instrumentation amplifier. However, the rejection ratio diminishes for frequencies higher than 10–50 Hz. Instrumentation amplifiers with a high 50 Hz common-mode rejection ratio are required for measurements of, for example, the electrical activity of the muscles (electromyography), the heart (electrocardiography) or the brain (electroencephalography). Such measurements would be made impossible by the large common 50 Hz hum of the power line frequency without CMRR.

A basic differential amplifier circuit is depicted in Fig. 3.40. If $A_0 \gg R_2/R_1$ and $R_i' \gg R_1//R_2 \gg R_o'$, the output voltage of this amplifier is given by:

$$v_o \approx v_1 \frac{R_2'(R_1 + R_2)}{R_1(R_1' + R_2')} - v_2 \frac{R_2}{R_1}$$

This expression can be found by first setting v_2 to zero and calculating the contribution of v_1 to the output voltage v_o. The transfer function of v_1 to v_o can be thought of as determined by the voltage divider R_1', R_2' followed by the non-inverting voltage amplifier of Fig. 3.38(b). Subsequently v_1 is set to zero and the contribution of v_2 to v_o is calculated. The transfer of this configuration is equal to the gain of the inverting amplifier of Fig. 3.38(a). The total output voltage v_o equal to the sum of the contributions of v_1 and v_2. If we choose the four resistors to satisfy the following relation:

$$\frac{R_2}{R_1} = \frac{R_2'}{R_1'}$$

Figure 3.40. Differential amplifier circuit.

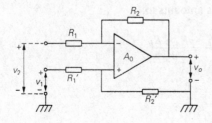

the output is zero when $v_1 = v_2 = v_c$. In that case there is no potential difference v_d between the input terminals, only a common-mode voltage v_c. In this case, the resistors R_1, R_2, R_1' and R_2' form as it were a balanced bridge with a supply voltage v_c and an output voltage equal to the output voltage of the amplifier. Since now $v_o = 0$, the right hand side of R_2 is also at ground potential. This situation is stable, because the bridge is balanced and the input signal of the operational amplifier is zero. When the resistance ratios on each side of the amplifier are equal, the differential gain is given by:

$$A_d = \frac{v_o}{v_1 - v_2} = \frac{R_2}{R_1} = \frac{R_2'}{R_1'}$$

A major disadvantage of the differential voltage amplifier configuration depicted in Fig. 3.40 is its low input impedance, which may load the measurement object considerably. This problem can be solved by incorporating two extra voltage amplifiers which compensate the input voltage (as in Fig. 3.38(b)), thus providing a very high input impedance. We then arrive at the circuit configuration of Fig. 3.41. If we first ignore the dotted cross at point A (the centre terminal is still grounded), the input stages are equivalent to the non-inverting voltage amplifier of Fig. 3.38(b). Their outputs are connected to a differential amplifier. In the input stages a common-mode signal is amplified as much as a differential-mode signal. All we have achieved here is a higher input impedance, but the common-mode rejection ratio is still the same. This can be improved considerably by disconnecting the ground at point A. In this situation, the gain of both input stages for a common-mode signal is reduced to unity. So when $v_1 = v_2 = v_c$, then also $v_1' = v_2' = v_c$. The magnitude of the differential gain can be calculated by assuming $v_1 = +v_d/2$ and $v_2 = -v_d/2$. The voltage across the input terminals of both operational amplifiers A_{01} and A_{02} is very nearly zero, due to large open-loop gain of the amplifiers and, consequently, the voltage across both resistors R_a is $v_1 - v_2 = v_d$. The current through these resistors equals $v_d/(2R_a)$. This current flows entirely through both resistors R_b, since the input current of the operational amplifier itself is negligible. Therefore, $v_1' = v_d/2 + v_d R_b/2R_a$ and likewise $v_2' = -v_d/2 - v_d R_b/2R_a$. The differential gain of the two input stages thus amounts to:

$$A_{d1} = \frac{v_1' - v_2'}{v_d} = \frac{R_a + R_b}{R_a}$$

We have already calculated the gain of the second stage, which we found to be given by:

$$A_{d2} = \frac{v_o}{v_1' - v_2'} = \frac{R_2}{R_1} = \frac{R_2'}{R_1'}$$

Therefore, the total gain is given by:

$$A_d = A_{d1} A_{d2} = \frac{R_a + R_b}{R_a} \frac{R_2}{R_1}$$

Remembering that the input stage gain A_{c1} for common-mode signals is equal to 1, we find for the overall common-mode rejection ratio:

$$\text{CMRR} = \frac{A_d}{A_c} = \frac{A_{d1} A_{d2}}{A_{c1} A_{c2}} = A_{d1} \frac{A_{d2}}{A_{c2}} = A_{d1} \text{CMRR}_2$$

Here CMRR_2 stands for the common-mode rejection ratio of the differential voltage amplifier in the second stage of this instrumentation amplifier. The rejection ratio is equal to the differential gain of the first stage, multiplied by the common-mode rejection ratio of the second stage. This is illustrated with the following typical values: if $R_a = 1$ kΩ and $R_b = 100$ kΩ, then $A_{d1} = 101$. If also $R_1 = 1$ kΩ and $R_2 = 100$ kΩ, then $A_{d2} = 100$. Since in practice R_2/R_1 will never be identical to R_2'/R_1', CMRR_2 will have a finite value. We can calculate this rejection ratio by assuming that the resistors have a tolerance better than 1%. The rejection ratio of the second stage will then exceed 2500. Thus, for the complete amplifier $A_d = 10\,100$ and $\text{CMRR} > 25 \times 10^4$.

In conclusion, the use of an input stage ahead of the differential amplifier increases the input impedance considerably, improves the differential gain and raises the common-mode rejection ratio.

The previous discussion assumes that the behaviour of the operational

Figure 3.41. Instrumentation amplifier with a separate differential input stage.

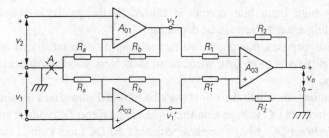

amplifiers is ideal (quasi-static). We ignored the fact that the gain A_0 depends on the frequency and that, therefore, we are dealing with a dynamic system. As the frequency increases, the open-loop gain A_0 decreases, reducing the feedback and increasing the inaccuracy. The phase shift of the amplifier can even cause the negative feedback to change into positive feedback for high frequencies. The amplifier may produce a spontaneous signal, i.e. oscillate, if no adequate measures are taken to prevent this from occurring.

Another effect we have not considered yet is that a practical operational amplifier will also exhibit static forms of non-idealness. One form, *offset*, is a form of *additive disturbance* which is the result of the bias current of the amplifying devices at the input of the operational amplifier. These devices cause a small current to flow in the input terminals, called the input bias current. Any polarity asymmetry within the input circuitry will also give rise to offset errors in the output voltage. From this output voltage an equivalent input voltage, which is referred to as the input referred *offset voltage*, can be calculated (simply by dividing by A_0). The offset voltage V_{off} and the input bias current I_{bias} are almost entirely determined by the type of amplifying device in the input stage of the operational amplifier. We have already mentioned that it is possible to compensate for these offset errors. However, V_{off} and I_{bias} also vary with temperature and, therefore, at a different temperature the compensation will not be complete and a residual offset will remain.

The values of the offset voltage V_{off} and the input bias current I_{bias} depend on the devices in the input stage of the operational amplifier. The offset voltage of integrated bipolar transistors is approximately 10 μV, with a temperature sensitivity of approximately 0.3 μV/K, and an input bias current of roughly 4 nA, with a temperature sensitivity of 25 pA/K. For integrated operational amplifiers with junction FET inputs, these figures are 250 μV, 8 μV/K and 30 pA, 3 pA/K, respectively. When, metal oxide semiconductor field-effect transistors (MOSFETs) are employed as input devices these numbers become 200 μV, 3 μV/K and 20 pA, 0.1 pA/K respectively. (The relatively high input bias current is caused by the protection circuitry safeguarding against electrostatic damage).

For most purposes, though, the performance of modern integrated operational amplifiers is more than adequate to make very good instrumentation amplifiers.

Applications for which the offset must be minimal sometimes require the use of an indirect DC voltage amplifier instead of a direct DC voltage amplifier. An *indirect DC voltage amplifier* converts the DC input voltage into an amplitude modulated signal by means of a non-linear operation (modulation). This AC signal is then amplified, after which it is demodulated into a DC signal again. Any remaining high-frequency signals are filtered out by an output low-pass filter. The modulator can achieve a lower V_{off} and I_{bias} than is possible with a direct DC voltage amplifier.

The most frequently used modulator is a switching modulator, which simply reverses the polarity of the input signal extremely rapidly. This type of modulator is referred to as a chopper modulator.

A chopper DC amplifier can be combined with a direct DC voltage amplifier to create a so-called chopper-stabilised amplifier (see Fig. 3.42). The direct DC voltage amplifier must have a large bandwidth, but does not necessarily have to exhibit excellent DC characteristics. In Fig. 3.42, the voltage V_{o2} represents the offset voltage of this amplifier A_{02}. Calculations show that the offset of the total configuration is reduced to V_{o2}/A_{01} by the presence of A_{01}. A low-pass filter is connected to the input of the chopper amplifier to avoid any possible high-frequency components of the signal from interfering with the chopper frequency. The high frequencies are only amplified by A_{02}, the low frequencies by both A_{02} and A_{01}. The chopper amplifier provides a much smaller offset error and a large open-loop amplification $A_{01}A_{02}$ for low frequencies. Therefore, when this amplifier is used in the feedback configuration shown in Fig. 3.42, great accuracy can be obtained for low frequencies.

An *instrumentation amplifier* (see Fig. 3.43) usually has an adjustable and accurate gain and often also an adjustable bandwidth. The instrumentation amplifier has a built-in power supply which is designed to suppress line interference and ground-loop injection of interference. The input amplifier is coupled to the rest of the amplifier by a transformer. In this way, the floating input stage can cope with very large common-mode voltages (up to 1 kV), without saturation of the amplifier. Often, the input stage is not really a

Figure 3.42. Chopper-stabilised amplifier configuration A_{01} is a low-frequency chopper amplifier, A_{02} is a DC coupled wide-band amplifier.

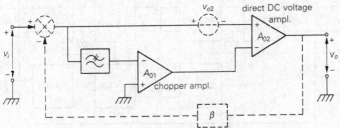

differential amplifier. The inputs of such a quasi-differential amplifier are marked 'Hi' and 'Lo', where the 'Lo' terminal is connected to the 'ground' of the floating input amplifier. The floating input amplifier is enclosed by a guard which is maintained at the common-mode potential, to avoid interference injection by the parasitic capacitance between the floating amplifier and ground. This was explained in detail in Section 2.3.3.3 (see Fig. 2.51).

Two examples of the specifications of instrumentation amplifiers are:

DC coupled instrumentation amplifier
- gain adjustable from 10^2 to 10^5, inaccuracy 10^{-4};
- temperature coefficient of the gain 10^{-5} K^{-1};
- static non-linearity $< 5 \times 10^{-6}$ of full range;
- input resistance 50 MΩ, output resistance 1 Ω;
- bandwidth adjustable from 0.01 Hz to 10 Hz;
- common-mode rejection ratio 10^7 (< 50 Hz);
- equivalent input noise (bandwidth 0.1 Hz): 2 nV;
- temperature sensitivity of amplifier null 1 nV/K.

Wide-band instrumentation amplifier
- gain adjustable from 1 to 1000, inaccuracy 10^{-4};
- temperature coefficient of the gain 3×10^{-5} K^{-1};
- static non-linearity $< 5 \times 10^{-5}$ of full range;
- distortion at 1 kHz $< 10^{-4}$;
- input impedance 10^8 Ω//100 pF, output resistance 0.1 Ω;
- bandwidth 100 kHz;
- settling time (within 1 part per thousand of final value): 80 μs;

Figure 3.43. Complete instrumentation amplifier.

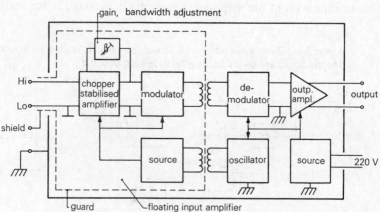

- common-mode rejection ratio 10^6 below 50 Hz, 10^4 below 1 kHz;
- temperature sensitivity of offset voltage 1 μV/K;
- temperature sensitivity of input bias current 0.5 nA/K.

It is pointless to increase the gain of a measurement system indefinitely. The sensitivity threshold of the system will soon be reached, due to noise and (additive) disturbances. This threshold can be shifted down somewhat by filtering. If the noise and disturbances are mainly *above* the frequency spectrum of the measurement signal, low-pass filtering is applied. If the disturbances are mainly *below* the frequency range of interest (for instance, line hum), *high-pass filtering* can be used to increase the sensitivity thres-

Figure 3.44. The effect of filtering a noisy signal. (a) The signal as measured. The signals (b), (c) and (d) are obtained by filtering signal (a) with different low-pass filters all of which have a –3 dB cut-off frequency of 100 kHz. The type of filter is different in each case: (b) elliptic filter (7 poles, 6 zeros); (c) Butterworth filter (fourth order); (d) Bessel filter (fourth order). Note the (linear) distortion of the impulse and the delay which is caused by the different filters. (Vertical: 100 mV/div, horizontal: 20 μs/div.)

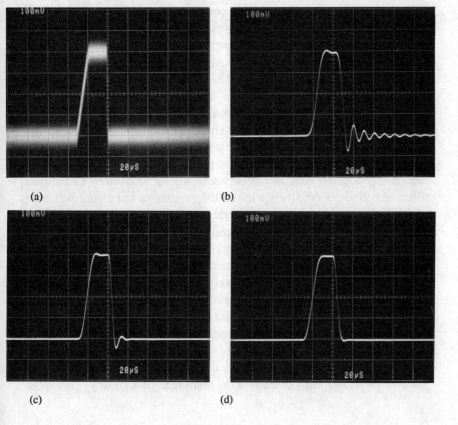

(a)   (b)

(c)   (d)

hold of the system. Both types of filtering can be combined in a *band-pass filter*. Finally, if the disturbing signals occupy a small band within the spectrum of the signal, a *notch filter* may be used to suppress a small part of the signal spectrum.

Filtering a signal can cause undesirable effects. This is illustrated in Fig. 3.44. The measured signal (a) contains noise. Most of the noise lies above the frequency spectrum of the signal and, therefore, the signal is filtered with a low-pass filter with a –3 dB point at 100 kHz. Figs. 3.44(b), (c) and (d) demonstrate the effectiveness of this; they contain almost no noise. However, it is clear that filtering has affected the trueness or fidelity of the waveform. Most of the filtered waveforms differ from that of (a), even though they are all obtained by filtering with low-pass filters with the same cut-off frequency. The differences are caused by the type of filter used. Another noticeable effect is the *delay* of the filters. The delay also depends on the type of filter used. Unfortunately, a detailed discussion of filters and their applications is not within the scope of this introduction to measurement and instrumentation.

As has been stated, filtering can considerably improve the sensitivity threshold of a measurement system (provided that it is performed correctly). In some cases, a much greater improvement can be achieved by the use of *synchronous* or *coherent detection*. An amplifier based on this detection principle will detect and 'lock onto' the phase of a signal. It is therefore called a *'lock-in' amplifier*.

The philosophy underlying the use of a synchronous detector in a measurement is as follows: Suppose our aim is a maximum useful sensitivity. The measurement is designed in such a way that the frequency spectrum of the measurement signal is as narrow as possible. The excess spectral width, which only contains noise and disturbances, can be removed by a band-pass filter. Applied *ad infinitum*, the resulting measurement signal will be a sinusoidal wave, band-passed through an extremely narrow filter. The necessary finite bandwidth will depend on the desired instrument reading time (response time). After the measurement signal is connected, or after an (abrupt) change, we must wait until transients in the signal have died out and a steady state is reached. The narrower the bandwidth, the longer this will take. In addition, the filter must also allow the modulation (i.e. the variation in time that contains the measurement information) of the signal to pass through. Modulation of a signal generates side-bands in the frequency spectrum, which must not be stopped by the filter, otherwise information is lost. So, bearing these limitations on the extent to which filtering can be used in mind, we must determine the amplitude of a (nearly) sinusoidal signal, with a known frequency. If we knew the phase of this signal (by having available a reference signal of the same frequency), all we would need to determine is the component of the measurement signal which remains in-phase with the reference signal. The noise and disturbances which are present within the bandwidth of the filter, but are not in phase with the reference signal, would not be detected. Even signal components which have a different phase than the reference are not detected. It is possible to prove that this method of detection is optimal (under the constraints given above).

A diagram of the way in which synchronous detection is applied is provided in Fig. 3.45. The figure shows that the lock-in amplifier (which performs the synchronous detection) requires a reference signal which contains the necessary frequency and phase information of the input signal. This reference signal is usually a sine or square wave, generated externally, and comes from or is supplied to the measurement object. If the measurement object does not generate this signal itself, it responds to this signal by producing an output signal which is a measure of the quantity being examined. This process of 'excitation' can also be applied to input transducers.

An illustration is given in Fig. 3.46. In order to measure the attenuation of a given narrow-band notch filter, a sinusoidal signal is applied to the input of the filter and also directly to a lock-in amplifier. This signal is used as an input signal as well as a reference signal. The output of the filter is very small (e.g. 1 V in, 30 μV out), due to the large attenuation in the bottom of the notch of the filter (e.g. −90 dB), and consists mostly of noise and distur-

Figure 3.45. Applying synchronous detection to a measurement situation to obtain the smallest possible sensitivity threshold.

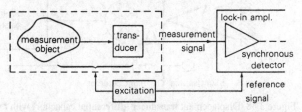

Figure 3.46. Measurement of the attenuation of a narrow-band notch filter.

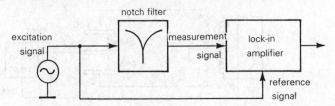

bances. Despite this, we can still retrieve the measurement signal, even from below the noise level, since the lock-in amplifier will only amplify the frequency component of that signal which is in phase with the reference; the rest of the signal will be rejected.

Fig. 3.47 shows another example of the application of a lock-in amplifier. The radiation thermometer (pyrometer) consists of an infra-red detector and a concave mirror which receives the infra-red radiation emitted by the measurement object. The radiation is 'chopped' by slots in a rotating disc. The back of the disc is coated with a reflective layer, so the detector 'sees' itself when the direct path to the measurement object is interrupted. In this way, the temperature of the detector jumps between a temperature which is a measure of the object surface temperature T_m and the (self-heating) temperature T_0 of the detector. The slots in the disc are hub-symmetrical, so the photodiode opposite the detector is illuminated simultaneously with the detector, thus providing a reference signal for the lock-in amplifier. The amplifier will only amplify the component of the small output signal of the thermal detector that is in phase with the reference signal. This component is a reliable measure of $T_m - T_0$, and is insensitive to disturbances.

Finally, Fig. 3.48 demonstrates the use of a lock-in amplifier in conjunction with a transducer. In this figure, the displacement Δx is measured with a differential capacitor, which is connected to a transformer in a bridge configuration. This configuration can perform extremely

Figure 3.47. A pyrometer with a lock-in amplifier for achieving a low sensitivity threshold.

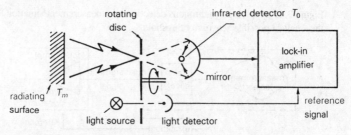

Figure 3.48. Displacement transducer (differential capacitor) with lock-in read-out circuit.

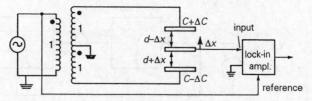

sensitive measurements, as all the noise, disturbances and distortion not in phase with the bridge excitation is filtered out.

We have seen that a lock-in amplifier must contain a synchronous detector. This detector can be implemented as a *product detector*, which multiplies the input signal $u_i(t) = \hat{u}_i \cos(\omega_i t + \phi)$ and the reference signal $u_r(t) = \hat{u}_r \cos(\omega_r t)$. The output signal is then filtered by a low-pass filter. The resulting signal is given by:

$$u_p(t) = \frac{1}{2} \hat{u}_i \, \hat{u}_r \, \{\cos(\omega_i t + \omega_r t + \phi) + \cos(\omega_i t - \omega_r t + \phi)\}$$

This result contains both the sum $(\omega_i + \omega_r)$ and the difference $(\omega_i - \omega_r)$ of the input signal and the reference signal frequencies. The sum of the frequencies is removed by the low-pass filter, so the remaining signal is given by:

$$u_o(t) = \frac{1}{2} \hat{u}_i \, \hat{u}_r \, \cos\{(\omega_i - \omega_r)t + \phi\}$$

In the case of *synchronous detection*, $\omega_r = \omega_i$. The output signal is then:

$$u_o(t) = \frac{1}{2} \hat{u}_i \, \hat{u}_r \, \cos \phi$$

As is shown in Fig. 3.49, a frequency band $2f_0$ from the input signal spectrum, located symmetrically around the reference frequency $f_r = \omega_r/2\pi$, is demodulated to the low-frequency region from 0 Hz to f_0 by the product detector and the low-pass filter (which has a -3 dB cut-off frequency f_0). The *bandwidth f_0* is chosen just large enough to include the *modulation $\hat{u}_i = \hat{u}_i(t)$*. This bandwidth also determines the *settling time*. To get a better

Figure 3.49. The principle of an analogue product detector: (a) block diagram; (b) spectrum of input signal; (c) spectrum of output signal. Note: B_n is the bandwidth of the band-pass filter which is used to remove noise and other interference from the input signal.

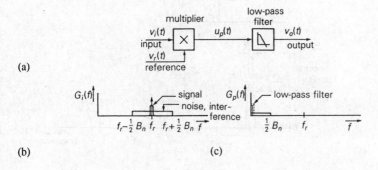

settling time one may opt for a larger than necessary bandwidth (but at the expense of a noisier response).

The output voltage of a product detector also depends on the phase. Usually, the phase difference between the input and reference signals is chosen to be zero for a maximum sensitivity.

The sensitivity of a product detector depends of the peak amplitude $\hat{u}_r$ of the reference voltage $u_r(t)$, which must therefore be determined accurately. In addition, the analogue multiplier used must not introduce any distortion, besides a pure multiplication of the two signals. Unfortunately, though, it is next to impossible to guarantee this over a large input amplitude range.

For measurement purposes, a different type of detector is most commonly used, i.e. the *switching detector*. Fig. 3.50 shows its principle of operation. The measured input voltage $u_i(t)$ is reversed at the rate of a square wave reference signal $u_r(t)$. This process can be described by multiplying $u_i(t) = \hat{u}_i\cos(\omega_i t + \phi)$ and a square wave signal with frequency ω_r and amplitude ± 1. In mathematical terms:

$$u_s(t) = \frac{4}{\pi} u_i(t) \,(\cos \omega_r t - \frac{1}{3} \cos 3\omega_r t + \frac{1}{5} \cos 5\omega_r t - ...)$$

Using $v_i(t) = \hat{u}_i \cos(\omega_i t + \phi)$, we find that now the output contains the sums and differences of ω_i and all the odd harmonics of ω_r. The amplitude of these frequency components is weighted by a factor $1/n$ for the nth harmonic of ω_r. If we low-pass filter this signal, the switching detector will only be sensitive to input signals which are located within a frequency band $2f_0$,

Figure 3.50. (a) The principle of a switching detector. (b) The corresponding spectral sensitivity of such a detector for the input signal.

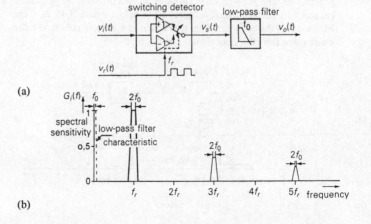

(a)

(b)

symmetrically around the odd harmonics of the reference frequency f_r. Thus, for synchronous detection (where $\omega_i = \omega_r$) using a switching detector, the output signal u_o of the nth harmonic $n\omega_i$ of the input signal is given by:

$$u_o(n) = \frac{2}{\pi n}\, \hat{u}_i(n) \cos \phi \quad (n \text{ odd})$$

In Fig. 3.50(b), the spectral sensitivity of a switching detector followed by a low-pass filter with a bandwidth f_0 is plotted. In order to eliminate the sensitivity to the higher harmonics in ω_r, a filter must be placed in front of the detector. This so-called *predetection filter* will reduce the signal energy of the higher harmonics (the first being $3\omega_r$) of the input signal to very nearly zero.

The major advantage of a switching detector is that it can be made to behave almost ideally over a large frequency and amplitude range. This large dynamic range is necessary for measuring signals which are drowned in noise and interference. Their much larger disturbance signals have to be dealt with linearly by the detector to exhibit the above described behaviour for the input signal. The fact that the detector needs an extra (predetection) filter is therefore taken for granted.

As mentioned before, a lock-in amplifier is an AC voltage amplifier which employs synchronous detection that allows one to measure the amplitude and phase of very small and noisy signals occupying only a relative narrow frequency band. The signal path $u_i(t)$ comprises a voltage amplifier AC and a predetection band-pass filter (see Fig. 3.51). Both the filter and the gain of the amplifier are adjustable. A section of the amplifier can be placed behind the filter to avoid saturation of the filter caused by noise and distortion. The reference path $u_r(t)$ includes adjustable amplification, a phase-shifter and a comparator to convert the reference into a square wave input for the switching detector. Phase-shifting is required, first of all, to ensure that $\phi = 0$ (maximum signal) exactly as the predetection filter will cause a phase shift of the input signal. Secondly, phase shifting enables us to measure the phase difference between the input voltage and the reference (zero signal when $\phi = 90°$). Often, the output stage also has an extra DC amplifier for post-amplification. When the input signal is extremely noisy, the total overall gain required is shifted towards the DC postamplifier, away from the AC preamplifier, to avoid detector saturation. For less noisy signals it is better to trade in the DC amplification for AC amplification to minimise drift in the output.

The following is an example showing how sensitive a measurement can be performed with a lock-in amplifier (see Fig. 3.48). We wish to determine the *fundamental* sensitivity threshold of a capacitive displacement sensor. This depends on the equivalent input noise of the input stage of the lock-in amplifier. (We will consider here only the noise and disturbances which are of a fundamental nature.) The input noise is characterised by a spectral noise voltage density (see Section 2.3.3.1) of approximately 5 nV/$\sqrt{\text{Hz}}$. The distance between the plates of the capacitor is $d \approx 1$ mm, the capacitance $C \approx$ 30 pF and the (sinusoidal) signal of the transformer is equal to 10 V, 100 kHz. If we limit the detection band to a width of 1 Hz, we find a sensitivity limit of 0.5 pm (5×10^{-13} m)! This means that a plate displacement of 0.5 pm would generate an output signal equal to the RMS-value of the noise at the output. Compare this to the distance between atoms in a crystal lattice, which is approximately 500 pm!

This sensitivity limit corresponds to a capacitance variation of 0.015 aF (1.5×10^{-20} F)! It will not be surprising that, in practice, we first encounter other disturbances which produce much larger effects. For instance, if we assume that the plates of the capacitor are made of 0.5 cm thick steel, the linear thermal expansion is approximately 55 nm/K, which is a factor of 10^5 larger! An air pressure variation of 1 kPa (10 mbar) results in a variation of the thickness of 50 pm. This is 100 times the fundamental sensitivity limit. Therefore, in practice, the smallest measurable displacement will depend entirely on the various mechanical limitations.

3.3.5 Non-linear signal conditioning

Non-linear signal conditioning is often used in measurement instrumentation to determine the amplitude characteristics of a periodic signal: the peak value, the average value or the RMS-value. It is also used for linearising any undesirable non-linear characteristics of transducers or other components of

Figure 3.51. Block diagram of a lock-in amplifier in which the signal path $u(t)$, the reference path $u_r(t)$, the synchronous detector and the output amplifier are shown.

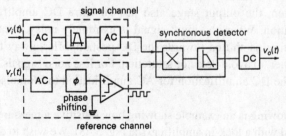

a measurement system. A non-linear transducer, for instance, can be corrected with a network of resistors and diodes if the non-linearity of this network is made to be the inverse function of the non-linearity of the transducer (series compensation).

In this section we will discuss several non-linear circuits, used to measure the peak value, the average value and the RMS-value of electrical signals.

Peak detectors

The peak value of an AC signal can easily be determined with a rectifying circuit. Fig. 3.52 illustrates this method of measuring the positive peak amplitude of an AC voltage. Such a peak detector is inexpensive and can easily be incorporated in a measurement probe. The cable between the probe and the DC voltmeter only carries a DC signal and the effect of the cable capacitance and the input capacitance of the measurement system is eliminated. Consequently, the predominant advantage of detecting the input voltage inside the probe is the increased input impedance. A peak detector probe can provide an input impedance Z_i of approximately 1 MΩ//1 pF. A disadvantage of peak detectors, shown in Fig. 3.52(b), is the non-linear behaviour for small input signals. This is a result of a knee in the VI characteristic of the forward biased diode. Therefore, this peak detector cannot really be used for voltages below 1 V, unless the scale of the voltmeter is made non-linear. This is not a very elegant solution though.

If the input signal contains a DC component, this is also detected by the peak detector, although this is often not desirable. The peak detector of Fig. 3.53(a) does not have this characteristic. The capacitor C_k will block any DC

Figure 3.52. Peak detector. (a) Positive peak detector using a single series diode. (b) Non-linearity of such a peak detector with a germanium diode and with a silicon diode.

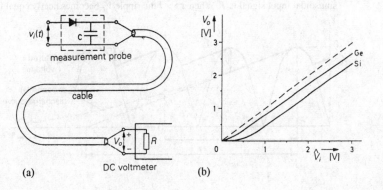

(a)

(b)

component in the input signal. Another peak detector circuit is given in Fig. 3.53(b), which in fact contains two peak detectors, one for the positive peaks and another for the negative peaks. Both peak detectors of Fig. 3.53 measure the peak-to-peak value, rather than the single-sided peak value.

The time constant $\tau = RC$ of the smoothing filter (the RC circuit which filters the DC component out of the rectified signal) must be much larger than the largest period T of the alternating signal. However, the peak detector must also be able to track any sudden variation in amplitude of the input signal, and therefore, the time constant may be no larger than t_0, if this is the time it takes the signal to change. This gives the requirement $t_0 \gg \tau \gg T$.

The ripple voltage V_r in the output of a single-sided rectifier is given by $V_r \approx \hat{V} \, T/\tau$. In the case of double-sided rectification this value is halved, so $V_r \approx \frac{1}{2} \hat{V} \, T/\tau$, provided that $\tau \gg T$. This can be seen from Fig. 3.54.

The non-linearity of a peak detector for small input signals can be improved considerably by applying compensation. Fig. 3.55 illustrates a signal peak value amplifier, which converts the peak $\hat{v}_i$ of an alternating

Figure 3.53. Peak detectors: (a) peak detector with a parallel and a series diode; (b) with two series diodes.

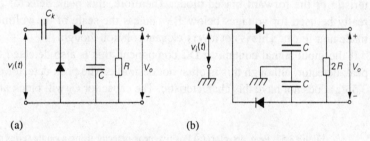

(a)　　　　　　　　　　　(b)

Figure 3.54. The ripple voltage V_r, superimposed on the output voltage V_o of a double-sided voltage rectifier, with a smoothing time constant τ. The period of the sinusoidal input signal is T. When $\tau \gg T$ the ripple V_r becomes nearly equal to V_r'.

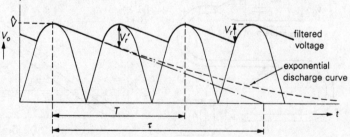

input voltage $v_i(t)$ into a much larger peak value $\hat{v}_o$ of the output voltage $v_o(t)$. The frequency f_o of the output signal is arbitrary and constant and is generated by an internal oscillator. The amplitude of the sinusoidal output oscillator signal is detected by a modulator and controlled to be equal to the amplitude of the output signal of the DC amplifier A_0. The latter amplifies the potential difference between the outputs of the peak detectors T_1 and T_2. One peak detector T_1 measures the input signal, whilst the other T_2 is supplied a fraction $R_1/(R_1 + R_2)$ of the AC output signal of the modulator. If the gain A_0 is made large, the feedback loop will ensure that the output voltage $v_o(t)$ is given an amplitude such that the output voltage of T_2 just compensates that of T_1. Assuming both detectors have identical non-linear characteristics (which are frequency independent), the input peak voltages of T_1 and T_2 must also be equal. Thus, the peak $\hat{v}_o$ of the output voltage must be a factor $(R_1 + R_2)/R_1$ larger than the peak $\hat{v}_i$ of the input voltage. If this gain factor is made sufficiently large, the peak value of the AC output signal will be much larger than the forward knee voltage of a diode. Therefore, it can easily be detected with a straightforward peak detector. If, for instance, $f_o = 100$ kHz and $(R_1 + R_2)/R_1 = 100$, a linear peak detector can be realised with a full-scale sensitivity of 10 mV and a frequency range from 100 kHz to 700 MHz, without the use of expensive wide-band amplifiers.

A peak detector is an inexpensive device for converting an AC voltage into a proportional DC voltage and it has a high input impedance. For this reason, peak detectors are sometimes used in AC voltmeters. The scale is usually calibrated to give the RMS-value of a pure sine wave input. So, if the peak value of a signal is equal to 1 V, the reading will be $\frac{1}{2}\sqrt{2} \approx 0.7$ V,

Figure 3.55. Peak value amplifier. The non-linear behaviour of T_1 is compensated by T_2 by means of a feedback loop. The transfer characteristic is determined by $\hat{v}_o = \hat{v}_i(R_1 + R_2)/R_1$.

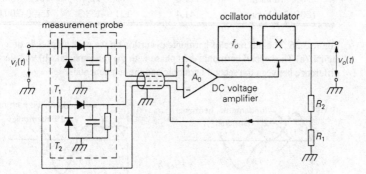

regardless of the input waveform. Consequently, if the waveform is not perfectly sinusoidal, such an AC voltmeter will display an incorrect value. The peak value of nearly sinusoidal signals depends strongly on the harmonic distortion of the signal. Fig. 3.56 shows two signals with the same RMS-value, but totally different peak values. This is due to the fact that the peak amplitude not only depends on the harmonic content, but also on the relative phase of the harmonics. Table 3.2 lists the variations of the peak, the average and the RMS-values when various amounts of second and third harmonic are added to the signal. The fundamental harmonic remains constant here for all cases. The phase difference of the higher harmonic and the fundamental is varied from 0° to 360°. The minimum and maximum values thus obtained are given in Table 3.2. We see that the average (of the absolute) amplitude of the AC voltage is affected less by the presence of higher harmonics than the peak amplitude.

Table 3.2. *The RMS amplitude, the average (of the absolute) amplitude and the peak amplitude of a fundamental harmonic (100%) plus various degrees of second or third harmonic.*

Signal	RMS %	Mean %	Peak %
Fundamental	100	100	100
with:			
10% 2nd harm.	100.5	100.0–100.5	90.0–110.0
20% 2nd harm.	102.0	100.0–101.9	80.0–120.0
50% 2nd harm.	111.8	100.0–110.1	75.0–150.0
100% 2nd harm.	141.4	125.0–129.9	112.5–200.0
with:			
10% 3rd harm.	100.5	96.7–103.3	90.0–110.0
20% 3rd harm.	102.0	93.3–106.7	87.1–120.0
50% 3rd harm.	111.8	89.9–116.7	107.6–150.0
100% 3rd harm.	141,4	121.9–133.3	154.0–200.0

Figure 3.56. The influence of harmonic distortion on the peak amplitude of a signal. (a) The second harmonic is in phase with the fundamental. (b) 90° phase difference between the second harmonic and the fundamental.

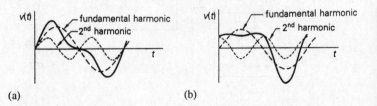

(a)   (b)

Average amplitude detectors

When one refers to the average value of an AC voltage signal, one is, in fact, referring to the average of the absolute value of the AC amplitude, or the average of the magnitude of the AC amplitude. This is precisely identical to the average of the double-rectified signal. This allows us to use the rectifier of Fig. 3.57. Here, the diode bridge will ensure that, regardless of the polarity of the input signal, the current through the meter will always flow in the same direction. The inertia of the moving coil meter will prevent it from vibrating at the rapidly alternating current repetition frequency. It will only respond to the average of the current. If the input voltage is connected directly to the bridge, the meter will exhibit a dead zone of twice the forward diode knee voltage ($2 \times (0.3$ to $0.8)$ V).

Due to this undesirable non-linear behaviour, the current through the meter will become zero when the input voltage is small. The current flowing in the meter coil will have a 'duty cycle' of less than 100%, hence the indicated value will lie below the true value.

It will be clear from Fig. 3.57 that the average magnitude of an input current will be measured correctly by the diode bridge. Therefore, the diode bridge is often included in the feedback loop of an amplifier as indicated in Fig. 3.58. If the gain A_0 of the operational amplifier is sufficiently large, the voltage across the resistor R will be approximately equal to the input voltage $v(t)$ and the current through R will amount to $v(t)/R$. Since the input impedance of the amplifier is extremely high, this current will be forced to flow entirely through the diode bridge. The current through the meter will then be given by $i(t) = |v(t)|/R$. Due to the inertia of the meter, it will only display the averaged value of this current which is a factor of $1/R$ larger than the input current. Thus, by applying feedback we have achieved two things:

Figure 3.57. Average amplitude magnitude detector. (a) Signal path for positive input voltages. (b) Signal path for negative input voltages. (c) Shape of input voltage $v(t)$ and the resulting current $i(t)$ through the coil of the meter.

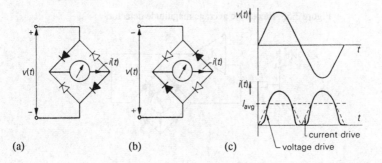

(a)　　　　　(b)　　　　　(c)

the input voltage is continually compensated (the input impedance is high, the measurement object is hardly loaded), the diode bridge is current driven, which eliminates the non-linear effect of the diode's knee voltage. An example of the specifications of such an electronic average magnitude meter with a feedback diode bridge is: measurement range: 100 μV–300 V full-scale, frequency range 10 Hz–5 MHz, inaccuracy 1%.

Voltmeters which measure the average magnitude of the input voltage are sometimes also calibrated in terms of RMS voltage. If the input is a pure sine wave $v(t) = \hat{v}\sin\omega t$, hence, $V_{\text{RMS}} = \frac{1}{2}\sqrt{2}\ \hat{v}$ and $V_{\text{avg}} = 2\hat{v}/\pi$. Therefore, for a sinusoidal input voltage with an RMS amplitude of 1 V, the meter would indicate an average magnitude of only 0.91 V. The meter scale is corrected for this; the meter will indicate 1 V for a sinusoidal input voltage. This meter will show an incorrect value when the input signal is not purely sinusoidal. However, the error resulting from (slightly) distorted input signals is not as large as that indicated by meters which rely on peak amplitude detection. A measure of the magnitude of the error introduced by the average magnitude indicating meter, when measuring the RMS-value, is given by the *form factor* of the input signal (see signal form sensitivity in Section 2.3.3.2).

RMS amplitude detectors

The RMS-value is a measure frequently used to characterise the amplitude of alternating signals. It is used for both deterministic signals and stochastic signals (i.e. noise). The RMS-value can be determined according to the definition: *the Root of the Mean of the Squares*. First, we must determine the square of the signal, then the average (for example, with a low-pass filter) and, finally, the square root of this result. The square and the square root of a signal can be obtained with the aid of a network of diodes and resistors; a so-called *function generator*. Such a function generator network is capable of realising non-linear operations on an input voltage signal V_i. The output signal V_o can be any arbitrary monotonic function of V_i. An example of one

Figure 3.58. Electronic average magnitude detector.

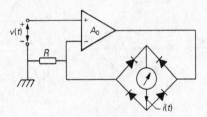

of these diode–resistor circuits is given in Fig. 3.59. This type of circuit is often also used for correcting non-linear transducers.

Provided that the gain A_0 is sufficiently large, the inverting input of the amplifier will act as a 'virtual ground'; the input impedance is virtually zero. The input voltage V_i is circuited across a number of parallel attenuator branches, formed by R_{aj} and R_{bj} (*j*th branch). As soon as the output voltage of one or more of the attenuators exceeds zero volts, a current I will flow to the inverting input of the amplifier. This is the case when $V_i > V_{ref} R_{aj} / R_{bj}$. Therefore, if we denote the threshold voltage at which the *j*th attenuator starts delivering current into the amplifier by V_j, we can write $V_j = V_{ref} R_{aj} / R_{bj}$. The total current I flowing into the virtual ground, when the first k diodes are conducting is given by:

$$I = \sum_{j=1}^{k} \frac{V_i - V_j}{R_{aj}}$$

Since the current I can only flow into the feedback resistor R, the output voltage is given by:

$$V_o = -R\,I = -R \sum_{j=1}^{k} \frac{V_i - V_j}{R_{aj}}$$

If the correct values are chosen for V_j and R_{aj}, the output voltage of V_o will approach a quadratic function of V_i (see Fig. 3.59(b)). The location of each break point of a subinterval of the function $V_o = f(V_i)$ depends on the resistors R_{aj} and R_{bj} and the magnitude of the reference voltage V_{ref}. Since the diodes start to conduct gradually, the relation between V_o and V_i will exhibit no discontinuities, but will be smoothed a little. It is also possible to

Figure 3.59. Forming the square of an input voltage signal V_i by means of a function generator network.

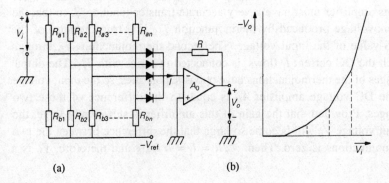

(a)　　　　　　　　　　　　　　　　　(b)

construct a function generator for negative voltages and even for non-monotonic functions.

Except for a quadratic function, it is also possible to realise the square root function with a similar function generator. However, if the RMS-value of a signal is determined in this manner, it will not be very accurate.

Another method of determining the RMS-value of a signal amplitude is given by the *heat definition*. This method is employed in a thermovoltmeter, in which the input current flows through a filament and heats it up. The generated heat is a direct measure of the RMS-value of the current. The temperature of the filament is measured with a thermocouple. To ensure a good thermal contact, the thermocouple is attached to the filament with a small glass bead, which also guarantees electrical insulation between the filament and the thermocouple. This configuration is encapsulated in a vacuum glass sphere to prevent loss of heat through convection. Therefore, the temperature of the filament can only decrease due to heat conduction and radiation. The temperature of the filament is made high to minimise the influence of the ambient temperature. However, this limits the robustness of a thermojunction to overloading. Also, a thermojunction has only a small efficiency (the ratio of the output power to the input power): about 0.1%. In addition, it responds only slowly (1–2 s). The sensitivity of a thermojunction can easily be calibrated with an accurately known DC current.

A thermojunction can be used for frequencies up to approximately 60 MHz. At higher frequencies, various parasitic effects (such as the inductance of the resistive filament, the capacitance between the heater filament and the thermocouple, etc.) will affect the correct operation. Over a frequency range from 10 Hz to 60 MHz the inaccuracy is less than 1%. Between 30 Hz and 10 MHz it may even be less than 0.1%.

The RMS-value of an AC voltage can be measured accurately by employing two identical thermojunctions in a configuration that compensates the output voltage, as shown in Fig. 3.60(b). With a feedback transconductance amplifier S, the input voltage V_{RMS} is converted into an output current I_{RMS}.

This amplifier must have a very accurate transfer function S, so that the thermovoltage produced by thermojunction Tj_1 is a true measure of the RMS-value of the input voltage. The second thermojunction Tj_2, through which the DC current I_f flows, is connected in series with Tj_1. The output voltages of the thermojunctions have opposite polarities, so the input voltage of the DC voltage amplifier A_0 is equal to the difference of these two voltages. Provided that the gain of this amplifier is sufficiently large, the output voltage V_o will become so large that the difference between the two thermojunctions is zero. Then: $V_o/R = I_f = \alpha I_{RMS}$ and therefore, V_o is a measure for the RMS-value of the input voltage. The accuracy of this electronic true RMS meter follows from the expression:

$$V_o = I_f R = \alpha I_{RMS} R = \alpha S V_{RMS} R$$

Here we have assumed that R is much larger than the resistance R_d of Tj_2. The factor α serves as a measure for the matching of Tj_1 and Tj_2 ($\alpha \approx 1$). S is the transfer function of the input stage: $S = I_{RMS}/V_{RMS}$. This shows that the absolute characteristics of Tj_1 and Tj_2 are no longer critical; it is only how well they are *matched* that matters.

Nowadays, thermojunctions are often replaced by integrated solid-state circuits. These consist of a differential amplifier and a pair of resistors. Both resistors are located very close to the base-emitter junction of the two input transistors of the differential amplifier. One resistor will carry I_{RMS}, whilst the other will carry the high-frequency measurement current I_f. Any unequal heating of the resistors will cause an offset of the differential amplifier. If the differential input transistor pair is made part of the operational amplifier A_0, then Tj_1, Tj_2 and A_0 of Fig. 3.60(b) can all be replaced by a single integrated circuit. This kind of RMS meter will operate easily at frequencies well above 100 MHz, since the parasitic impedances are far smaller due to the tiny dimensions of the circuit.

For measurements of the RMS-value of noise and of signals with a small 'duty cycle' we must be aware of the possibility of saturating the amplifier S. Although the RMS-value of a noise voltage or an impulse voltage may be small, it may still have very large peak value. If such signals are measured with electronic measurement systems it may happen that as a result the amplifiers are already saturating, while the read-out (the RMS-value) is still

Figure 3.60. (a) Thermojunction. (b) Electronic RMS amplitude detector.

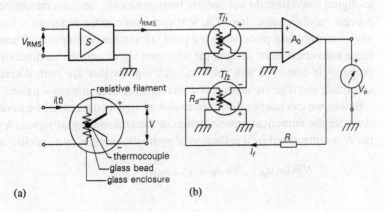

(a)   (b)

small. For these measurements, the so-called *crest-factor* is important. This is the ratio between the peak value and the RMS-value of a signal. Ordinary electronic RMS meters can mostly cope with signals with a crest-factor of up to 10.

3.3.6 Digital-to-analogue and analogue-to-digital conversion

A signal whose amplitude is time dependent is called a *dynamic signal*. The magnitude of a dynamic signal can be described as a function of time. If the function is defined for all points in time (over a certain finite time interval), we refer to this signal as a *continuous-time* signal. If the amplitude of the signal, and thus the function value, can assume any value within a certain amplitude interval, this signal is called a *continuous-amplitude* signal. Almost all signals which are generated by macroscopic physical processes are both continuous-time and continuous-amplitude signals. Such continuous signals which fluctuate in relation with (continuous) physical processes are therefore called *analogue signals*.

There are also *discrete-time* signals. The amplitude of such a signal is only available or known at certain, discrete points in time. We can consider a discrete-time signal as the result of the *sampling* of a continuous-time signal (see Section 2.2).

In the same vein, it is also possible that the amplitude of a signal can assume only certain, discrete values. Then the signal is referred to as a *discrete-amplitude* signal. Such a signal can only assume a finite number of amplitude values between given upper and lower limits. The process of converting continuous-amplitude signals into descrete-amplitude signals is called *quantisation* and is realised by an analogue-to-digital converter. Some of these converters operate instantaneously. Then the continuity in time is maintained even for the discrete-amplitude signal. However, most analogue-to-digital converters do not operate instantaneously, because the conversion process requires some time. The next conversion in the sequence of conversions can only take place when the prior conversion is completed. Therefore, these converters must sample the analogue signal and the continuous-time property is lost. We will refer to such signals that are both discrete in amplitude and discrete in time as *digital signals* (Lat.: *digitus* = finger).

Before we can continue this discussion of signal conversion, we must first consider the numerical representation of the ensuing digital signal. A number N is represented by a collection of symbols, arranged in a specific order:

$$N = (a_n a_{n-1} \dots a_1 a_0, a_{-1} \dots a_{-m})$$

In this expression n and m are integers. If we take a numerical representation, this representation means:

$$N = a_n r^n + a_{n-1} r^{n-1} + \ldots a_1 r^1 + a_0 r^0 + a_{-1} r^{-1} + \ldots + a_{-m} r^{-m}$$

in which r is the radix or base of the numerical representation; r is an integer larger than 1. The coefficients a_i are integers, for which $0 \le a_i \le r - 1$.

Popular numerical representations are based on the radix 10, the *decimal* system; base 8, the *octal* system; base 3, the ternary system and base 2, the binary system. For instance, in the decimal system 701.43 means $7 \times 10^2 + 0 \times 10^1 + 1 \times 10^0 + 4 \times 10^{-1} + 3 \times 10^{-2}$. In the binary system $1011.01 = 1 \times 2^3 + 0 \times 2^2 + 1 \times 2^1 + 1 \times 2^0 + 0 \times 2^{-1} + 1 \times 2^{-2}$ which, in decimal notation, is equal to 11.25.

In the decimal system ($r = 10$) the coefficients a_i are referred to as decimals, whereas in the binary system ($r = 2$) they are simply called *binary digits* usually compressed to *bits*. The first bit a_n is the Most Significant Bit (MSB) and the last bit with the smallest value is the Least Significant Bit (LSB). Usually a digital signal is represented in the form of a sequence of binary numbers; a binary signal. The two distinguishable levels in such a binary signal are often denoted by '0' and '1'. The '0' value usually corresponds with the smaller amplitude of the signal and the '1' value to the larger.

The preference for the binary system is a consequence of the fact that there are many simple electronic, hydraulic, and other circuits which have two stable states, such as a switch (open or closed), a relay, a flip-flop and a ferrite magnetic memory ring core (two directions of magnetisation). Computers therefore also use the binary system. To enable computers to work with analogue quantities, these must be first converted into bit strings with analogue-to-digital converters. The reverse operation is required, of course, when a computed result must be available in analogue form, for instance, in a process controlled by an analogue quantity such as a current.

An observer reading the position of a pointer placed in front of a dial is in fact also an analogue-to-digital converter; discrete values are assigned to an analogue deflection.

Digital-to-analogue conversion

A Digital-to-Analogue Converter (DAC) maps a digital signal into a analogue signal. For the sake of simplicity, we will only consider analogue voltage signals in the following. The digital signal is represented by n parallel bits which, at a given point in time, have the value:

$$D = (a_n a_{n-1} \ldots a_1 a_0)$$

If, for the time being, we disregard fractions and the sign, the DA conversion is simply described by:

$$V_A = V_0 D = V_0 \sum_{i=0}^{n} a_i 2^i$$

in which V_0 is a small, fixed incremental voltage and V_A is the analogue output voltage of the DAC. The transfer characteristic of the DAC is plotted in Fig. 3.61(a). D can only assume a finite number of discrete values. The minimum step with which V_A can increase is precisely equal to V_0.

Fig. 3.61(b) shows a simple realisation of a DAC. The digital input D consists of all bits a_i ($i = 0, 1, \ldots, n$). If $a_i = 1$, the corresponding switch a_i is connected to a negative reference voltage $-V_R$; if $a_i = 0$, the switch is connected to ground. As the open-loop gain A_0 of the operational amplifier is extremely large, the node S is virtually grounded. The input current I of this current-to-voltage amplifier is the sum of the currents of all resistors R_i for which $a_i = 1$. This current is given by:

$$I = -\sum_{i=1}^{n} \frac{a_i V_R}{R_i}$$

The analogue output voltage now is $V_A = -I R_t$. In order that $V_A = V_0 D$ we choose the value of resistor R_i to be equal to twice that of the following resistor R_{i+1}. So, if $R_0 = R$, then $R_1 = R/2$, $R_2 = R/4$ and $R_i = 2^{-i}R$. Then we find for V_A:

Figure 3.61. (a) Transfer characteristic of a DAC. (b) Realisation of a DAC in which each resistor value R_i corresponds to a binary digit or bit.

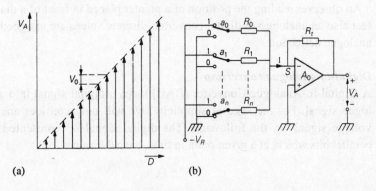

(a)   (b)

$$V_A = \frac{V_R R_t}{R} \sum_{i=0}^{n} a_i 2^i = V_0 \sum_{i=0}^{n} a_i 2^i$$

For an 11-bit converter, $n = 9$. With $V_R = 5$ V, $R_f = 1$ kΩ and $R_0 = R = 1$ MΩ, the smallest incremental voltage step is $V_0 = 5$ mV. The smallest resistor R_{10}, which corresponds to the MSB a_{10} must be 2^{10} smaller than R_0, so $R_{10} = 1/1024$ MΩ. (Keep in mind here that n is the subscript of the switch bits a_i: $i = 0, 1, 2, \ldots , n$. So there are $n + 1$ bits!)

A major disadvantage of this DAC is the fact that the ratio between the largest (R_0) and the smallest (R_n) resistors becomes unpractically large for a large number of bits. If the number of bits is n, this ratio is equal to $R_0/R_n = 2^n$. When the dimensions of the resistors are small (for instance, in thin-film processes) it is difficult to realise such a wide range of resistance accurately. The smallest resistors R_n, R_{n-1}, which belong to the most significant bits a_n, a_{n-1}, especially must have an accurately determined ratio to the feedback resistor R_f.

The resolution of the DAC is equal to the maximum output voltage divided by the incremental voltage step V_0. Thus, the resolution r is:

$$r = \sum_{i=0}^{n} 2^i = 2^{n+1} - 1$$

Obviously, the resolution of a DAC must be high enough to implement the requirements for the accuracy of the measurement system. If the resistors of the DAC all were to have extremely accurate values, but the resolution were low, the inaccuracy of the DAC would be determined entirely by the large quantisation error, resulting from the low resolution. Therefore, if the allowable relative inaccuracy in the expression $V_A = V_0 D$ is denoted as ε, the resolution must satisfy $r \geq 1/\varepsilon$.

The inaccuracy of the DAC of Fig. 3.61(b) can reach approximately 10^{-3} based on the resistance values above. This requires a resolution of 10^3, i.e. at least 10 bits, since $2^{10} - 1 = 1023$.

A type of DAC which contains only two different resistors is illustrated in Fig. 3.62. The circuit consists of a resistive ladder network and a current-to-voltage amplifier. It is evident that the internal resistance of the sections of the ladder to the right of the nodes 0, 1, 2, ... n remains the same $(2R)$, regardless of the node, since the switches are always at ground potential (in either setting $a_i = 0$, or $a_i = 1$). As a consequence, the current at each node 0, 1, 2, ... n will divide equally between this internal resistance $2R$ to the right of the node and the resistor $2R$ connected to the switch. The current which flows from the reference source V_R to node n is equal to $-V_R/R$. At this point it is divided equally into two currents $-V_R/2R$, one flowing to the switch associated with bit a_n and one to the rest of the ladder to right of this node. This process is repeated at node $n - 1$, so the current through the resistor and switch associated with bit a_{n-1} becomes $-V_R/4R$ etc. These currents are summed at the virtually grounded input of the operational amplifier. The current corresponding to the LSB a_0 is $-V_R/2^{n+1}R$, which is exactly equal to the minimum possible voltage step V_0 of the output. We find for this minimum step $V_0 = V_R/2^n$ and for the total transfer function of the DAC:

$$V_A = -2R \sum_{i=0}^{n} \frac{-a_i V_R}{2R \, 2^{n-i}} = V_0 \sum_{i=0}^{n} a_i 2^i$$

This configuration requires that the *ratio* of the resistances R and $2R$ of the ladder resistors associated with the highest bits a_n, a_{n-1}, ... and the feedback resistance $2R$ is accurately defined. As only two values of resistance are being used, this DAC can achieve a far greater accuracy than the previous one. An inaccuracy below even 10^{-4} is possible. Then, though, we must use at least 14-bit resolution, since $2^{14} = 16\,384$.

Analogue-to-digital conversion

In mathematical terms, analogue-to-digital conversion is described by:

$$D = \sum_{i=0}^{n} a_i 2^i = \frac{V_A}{V_0} + Q$$

Here V_A is the input voltage to be converted, V_0 the minimum voltage step that can be resolved, and Q is the remainder which arises from the quantisation error. The task of an Analogue-to-Digital Converter (ADC) is to find values for the bits a_i for which the quantisation error Q is minimal. Clearly,

Figure 3.62. DA conversion using a resistor ladder network.

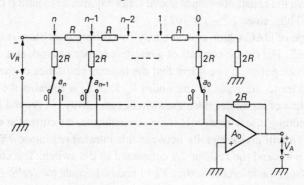

one always expects a non-zero quantisation error: the voltage V_A is a continuous-amplitude signal, whereas D can only assume discrete values (see Fig. 3.63(a)).

The most accurate ADCs are realised by employing automatic compensation (see Section 3.3.2). This can be accomplished with a DAC in a feedback loop arranged such that the DAC output voltage compensates the input voltage V_A. If the input voltage is continuously compensated, the resulting ADC is called a *following* or *servo-type ADC* (see Fig. 3.64(a)).

The input voltage may also be compensated only at given, discrete points in time. This type of ADC has to sample the input signal. The moment of sampling is usually given by a conversion start pulse. An example of such an ADC is the *successive-approximation ADC*.

Let us examine the servo ADC in more detail. Assuming that the (binary) counter in Fig. 3.64(a) is preset to zero immediately after the ADC is

Figure 3.63. (a) Transfer characteristic of an ADC. (b) The quantisation error Q, which is inherently associated with the ADC.

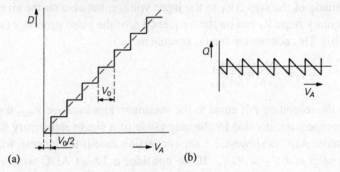

Figure 3.64. (a) Example of a servo ADC. (b) The compensation voltage V_c as a function of time, immediately after power-on.

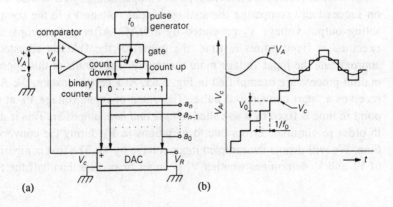

switched on, the output voltage V_c of the DAC will be zero. Therefore, the comparator, which compares V_c to V_A, will receive a positive difference voltage $V_d = V_A - V_c$. The output of the comparator will switch the counter into the 'count up' mode, at the rate of the pulse generator. The voltage V_c will increase in steps. The width of these steps in time is equal to the period of the pulse generator. At a certain point, V_c will exceed the input voltage V_A. Now V_d becomes negative, the output of the comparator switches and the counter starts to count down. Therefore, V_c decreases and V_d becomes positive, etc. The result is that V_c will fluctuate step-wise around V_A. The magnitude of the resulting quantisation error depends on the resolution of the internal DAC. The digital output signal $D = a_n a_{n-1} \ldots a_0$ is equal to the value of the counter, which generates the binary inputs of the DAC.

A servo ADC can follow small variations in the input signal V_A relatively quickly. However, if V_A exhibits a large step variation, the ADC will no longer follow it immediately, since it can only gradually approach the new value of V_A in many small constant magnitude steps (staircase function). The time necessary for reaching the new value of V_A depends not only on the magnitude of the step ΔV_A in the input voltage, but also on the size of the elementary steps V_0 and on the frequency f_0 of the pulse generator (see Fig. 3.64(b)). This conversion time t_c is equal to:

$$t_c = \frac{\Delta V_A}{f_0 V_0} = \frac{r \, \Delta V_A}{f_0 V_{\max}}$$

since the resolution r is equal to the maximum input voltage $V_{\max}$ the ADC can compensate, divided by the magnitude of a single elementary step V_0. The servo ADC is slowest, i.e. the conversion time is the longest, when the input step is $\Delta V_A = V_{\max}$. If we consider a 12-bit ADC with a clock frequency of $f_0 = 200$ kHz, the resolution is $r = 2^{12}$ and $t_c = 20.48$ ms for $\Delta V_A = V_{\max}$.

The successive-approximation-type of compensating ADC is also based on successively comparing the analogue input voltage V_A to the compensating output voltage V_c generated by a DAC. After each comparison, executed at fixed points in time, the output of the DAC is adjusted to approximate the input voltage more accurately. The details of this approximation process are exemplified in Fig. 3.65. At a certain instant, the ADC receives a 'start conversion' pulse. The value of input voltage V_A at that point in time is fixed by a so-called *sample and hold* amplifier. This is done in order to eliminate errors due to variations in V_A during the conversion time. We will denote the sampled input voltage by V_A'. The first comparison of V_A' and V_c determines whether V_A' is larger or smaller than half the full-scale value V_{max}. In the example illustrated, the result is 'larger than' and so the MSB a_n is set to 1. The following comparison, a fixed time interval later, determines whether V_A' is larger or smaller than 3/4 of V_{max}. In the example, again, the result is 'larger than', so also $a_{n-1} = 1$. In this manner, the compensation voltage V_c is changed at the ith comparison by a step value 2^{-i} V_{max}. If the result of the previous comparison sets the corresponding bit to 1, the following step is positive, if the bit is set to 0 the following step is negative. If the required resolution is r, then $r = n + 1$ comparisons must be performed to approximate V_A' to within $\pm 2^{-r} V_{max}$ volt.

After these $n + 1$ comparisons, the conversion is completed and the ADC will wait for a new start pulse. The conversion time t_c is constant here and given by $t_c = (n + 1)/f_0$ if f_0 is the number of comparisons per second.

When $f_0 = 200$ kHz and $r = 12$ bits, the conversion time $t_c = 60$ μs. Note that this value is independent of the magnitude of the input voltage step. Therefore, for signals which exhibit large step variations, successive-approximation ADCs are faster than servo ADCs. If the signal varies only gradually, a servo ADC will be faster.

We will discuss the circuitry necessary for making a successive-approximation ADC with the aid of Figs. 3.66 and 3.67. We will consider here the ADC without internal feedback, used in the previous examples; we will only discuss open-loop ADCs. Fig. 3.66 shows such an ADC without a feedback loop. The input voltage V_A is fixed with a sample-and-hold circuit during the time it takes to convert the input into a digital output. The voltage V_A' is compared to half of the full-scale value V_{max} and the value for bit a_n is found. If $a_n = 1$ a voltage equal to $\frac{1}{2} V_{max}$ is subtracted from V_A', if $a_n = 0$ V_A' is passed on to the next cell unchanged. The result of the first cell is supplied to the next cell. Here it is compared to $\frac{1}{4} V_{max}$ and the value of bit a_{n-1} is

Figure 3.65. Compensation process in a successive approximation ADC.

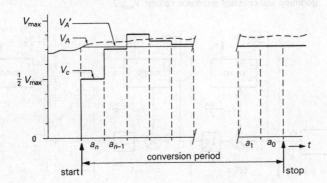

determined. This procedure is continued until all $n + 1$ bits have been determined. The $n + 1$ cells are all identical; the only difference is that the *reference voltage* is *halved* in each subsequent cell. Therefore, the voltage level at which the comparators operate becomes successively smaller. This is accompanied by increasing problems with offset, drift, noise, etc.

For this reason another, better, method exists, based on *doubling* of the voltage to be measured. This is shown schematically in Fig. 3.67. The first cell, which calculates bit a_n operates in exactly the same way as in the previous case. However, the remaining voltage is now amplified by a factor of 2 and the reference voltage level $V_{max}/2$ is kept the same for all cells. This means that the comparators operate at nearly the same voltage levels for all $n + 1$ cells. Instead of using $n + 1$ identical cells, often a single cell with two analogue (capacitive) memories is used. One memory capacitor samples and holds the input voltage of the converter cell and the other capacitor holds the remainder voltage (cell output voltage). By using switches to exchange these

Figure 3.66. A successive approximation ADC without internal feedback which makes use of halving the reference voltage in each subsequent cell.

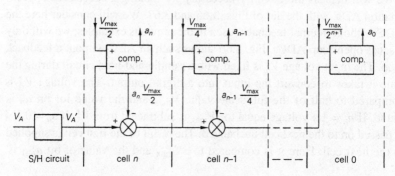

Figure 3.67. Open-loop successive approximation ADC with (remainder) voltage doubling and constant reference voltage $V_{max}/2$.

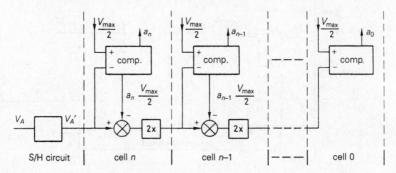

two memories, the same cell can be used sequentially $n + 1$ times. This method is referred to as '*recycling of the remainder*'.

More and more applications of ADCs require a high-speed operation. The fastest possible conversion method is indicated in Fig. 3.68. The input voltage V_A is compared simultaneously to a large number of different reference voltages. A logic circuit converts the comparator outputs into the binary representation $D = (a_n a_{n-1} \ldots a_0)$. Evidently, for an $(n + 1)$-bit converter, such a 'flash' converter requires $2^{n+1} - 1$ comparators. The method is fast, but also very costly. A quick calculation shows that for an 8-bit converter, no less than 255 converters are required! Therefore, to keep the number of comparators within reasonable limits, the circuit of Fig. 3.69 is often used instead. The basic principle is straightforward. An 8-bit converter will require two 4-bit flash bit converters, each containing only 15 comparators. The upper ADC of Fig. 3.69 provides only a coarse conversion giving the 4 most-significant bits. The voltage which corresponds to the

Figure 3.68. A so-called flash ADC (parallel converter).

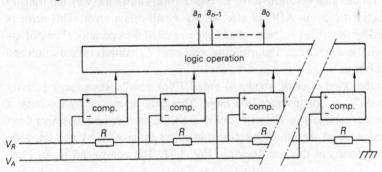

Figure 3.69. The principle of a so-called 'dual rank' or 'pipelined' ADC.

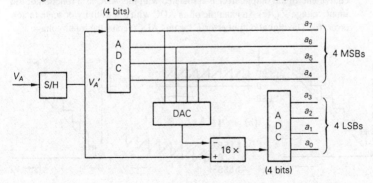

value of these 4 bits is created by a DAC and subtracted out from the sampled input voltage V_A' . The remainder voltage, which is a measure for the large quantisation error of the first DAC, is amplified $2^4 = 16$ times and supplied to an ADC identical to the first one. This second ADC produces the 4 least significant bits of the 8-bit converter. The size of this ADC has been reduced considerably compared to a full flash converter (from 255 to 2×15 comparators), at the expense of (a reduction in conversion) speed.

The previous discussion only covers the most important types of DAC and ADC. There are many more methods of realisation for AD conversion. In Section 4.3 we will discuss several forms of *integrating AD converters*, as these are frequently used in digital voltmeters mainly because they give good interference suppression of line hum.

In Fig. 3.63 it was shown that all ADCs produce quantisation errors, which are inherent to all signal conversions with a finite amplitude resolution. However, there are other errors which should be taken into consideration as well.

In Fig. 3.70 two types of error are shown together with the effect they have on the total resulting error E. Fig. 3.70(a) shows the way the transfer characteristic of an ADC is affected by a null-offset error. This error is caused by the offset voltages and bias currents of the operational amplifiers used in the converters. The resulting total error E, exhibits quantisation and offset errors.

Another kind of error, shown in Fig. 3.70(c), results when the sensitivity of an ADC is, for instance, too large. This may be caused by too large a reference voltage V_R. The effect of this on the total error E has also been plotted. One of the most objectionable errors in an ADC is non-linearity. Several forms of this are drawn in Fig. 3.71. The non-linearities of Figs.

Figure 3.70. (a) The transfer characteristic of an ADC with an offset error. (b) The equivalent digital output error E associated with this ADC as a function of the input voltage V_A. (c) An example of an ADC with a sensitivity or scale factor error. (d) The digital output error E for this ADC versus the input voltage V_a.

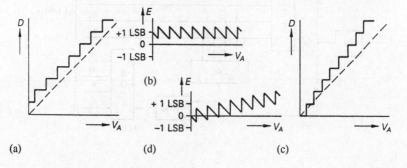

3.71(a) and (b) occur when the resistors, which determine the values of the various bits, deviate from their nominal values. A very serious form of non-linearity is shown in Figs. 3.71(c) and (d). Here, the deviation of the resistor values is so large (or one of the switches could be malfunctioning), that one or more of the possible digital outputs is completely skipped. This is sometimes called a 'missing-bit' error. Also shown at the top of the characteristic in Fig. 3.71(d) is a so-called monotonicity error. This error is especially disruptive in a feedback system. At first, when V_A increases, D also increases, but at a certain point around this error, D starts to decrease again. This results in a change of the polarity of the feedback (a 180° phase shift) turning negative feedback into positive feedback. This kind of excessive non-linearity can cause instability in controlled systems.

3.4 Measurement displays

The presentation of the measurement result to a human observer is performed by a (measurement) display. Humans can only acquire information by means of their senses. We, humans, are not very good at directly observing electrical signals with any reasonable accuracy, let alone dynamically varying electrical signals. Therefore, for presenting measurement information, we need transducers which can convert electrical signals into objectively observable signals. For measurement purposes usually the only sense that we use is sight, as this can easily handle a relatively large amount of information correctly. Sometimes, though, hearing is used for detecting signals accompanying the measurement, such as warnings, alarms, etc. In the past, hearing was sometimes used for null detection. The bridge was

Figure 3.71(a) An ADC with a non-linear transfer characteristic. (b) The associated digital output error E. (c) An ADC with missing bits and a non-monotonic transfer. (d) The output error of this ADC.

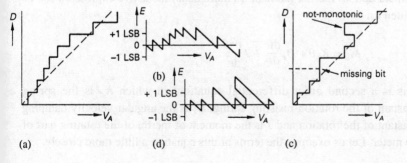

balanced with the aid of a pair of headphones.

Transducers which convert electrical signals into a visual signal are called *displays*.

As we have seen in Section 2.3.3.4, we can distinguish between *analogue* or *continuous displays* and *digital* or *discrete displays*. Digital displays are usually numerical displays; the measured quantity is presented in the form of numbers. However, this is not always necessary, for instance, for a go, no-go display associated with a quality test in a production line. The more different states a discrete display has, the better the resolution and the more detailed the representation.

In the following, we will restrict ourselves mainly to analogue displays. Most visual analogue displays are either electro-mechanical or electro-optical devices. *Electro-mechanical displays* convert an electrical signal into a proportional analogue mechanical signal. This mechanical signal is normally the translation or rotation of a moving mechanical system (pointer) with respect to a fixed mechanical reference (scale, dial). These electro-mechanical displays are usually referred to as 'meters'.

In contrast to mechanical displays, which must be illuminated externally in order to be read, electro-optical displays convert an electrical signal into a (visible) light signal. Examples of electro-optical displays are: the cathode-ray tube of an oscilloscope and the numerous applications of LEDs.

3.4.1 Electro-mechanical displays

We saw in Section 2.3.3.2 that the equation of motion of an electro-mechanical display or 'meter', converting an electrical signal into a proportional mechanical rotation, is given by:

$$M_d = M_r + M_{da} + M_i$$

in which M_d is the driving moment exerted on the rotating armature and the pointer of the display, M_r the restoring moment, M_{da} the angular damping moment and M_i inertial moment. In more detail the above expression can be written as:

$$M_d = K_r\theta + D_r\frac{d\theta}{dt} + J\frac{d^2\theta}{dt^2}$$

This is a second order differential equation, in which K_c is the spring constant of the rotation-resisting spring, D_r is the angular velocity damping constant of the rotation and J is the moment of inertia of the rotating part of the meter. Let us examine the terms of this equation a little more closely.

In the steady state, in which the pointer is at rest, the equation of motion reduces to $M_d = M_r = K_r\theta$. The accuracy of the angular deflection θ of the pointer as a measure for the driving moment (which, in turn, is a measure for the input quantity) thus depends on the spring of the meter. For this reason, the material and the shape of the spring are chosen carefully to ensure that it is linear over the entire range of θ, so that K_r is independent of θ. A common spring material for this application is phosphor bronze and the spring is often given the shape of a spiral or suspension strip (torsion spring). The rotating armature and pointer of a meter with a spiral spring are usually mounted on trunnion bearings. When a torsion spring is used, the rotating part is suspended from the elastic strip spring. This has the advantage that there is no wear and virtually no friction. Also, there is no dead zone due to stiction, as with trunnion bearings.

The second term in the right-hand side of the equation above: M_{da} ($= D_r\omega$) expresses the moment of the torque which results from the velocity dependent damping of the moving part. This air drag is often increased by a small vane connected to the hub of the meter. The damping moment from the air resistance of the vane can be described by a series expansion:

$$M_{da} = D_1\omega + D_2\omega^2 + D_3\omega^3 \dots$$

If the shape of the paravane is chosen correctly, the higher order terms may be neglected, provided that the angular velocity ω is not too large. In this manner, *air damping* provides a damping moment which varies linearly with the angular velocity ω. As an alternative, electro-magnetic damping is sometimes used instead. This damping is realised by attaching a thin conducting disc (usually aluminium) to the movement of the meter, exposing the disc to a large permanent magnetic field. As the disc moves in this field, eddy currents are induced, which produce a resisting moment proportional to the angular velocity ω of the disc. If there is already a permanent magnet present in the meter for generating the driving moment M_d, such as is the case in moving coil meters, the pointer may also be damped by short-circuiting one single winding of the coil. This method also works when the coil is wound on a conducting (aluminium) bobbin. When the bobbin moves in the magnetic field, a current is induced, which produces a moment proportional to its angular velocity. If the magnetic field has radial symmetry, the moment will be independent of the pointer angle θ.

The last term $M_i = J\, d^2\theta/dt^2$ is the source of the second order behaviour of the equation of motion that governs the meter. It is important to minimise the moment of inertia J, especially when the meter is designed for a quick response. Therefore, the mass m of the moving part must be small.

Remembering that the moment of inertia of a mass Δm at a distance r from its axis of rotation is equal to $r^2\Delta m$, we also conclude that the radial dimensions of the armature must be kept small. Therefore, for a fast display, an optical pointer rather than a mechanical pointer is used. Such an optical pointer consists of a small mirror, mounted on the axle of the meter, which deflects an incident beam of light over an angle 2θ, when the mirror rotates across an angle θ. The further away the scale on which the light beam is made visible, the higher the sensitivity (angular magnification). In this way, J can be reduced considerably. When the light beam is reflected several times by a number of mirrors the size of the instrument can be kept small (folded beam). Another advantage of this optical pointer is that it does not give rise to parallax errors.

Due to the inertia J of the moving part of the meter, it is not capable of following rapid fluctuations of the measured quantity. This (low-pass filter) characteristic is often used for determining the average value of a quickly fluctuating quantity. The driving moment M_d then varies so quickly, that the pointer of the meter will remain at an angle $\theta = \theta_{\mathrm{avg}}$. Therefore, the time derivative of θ in the differential equation will be zero, resulting in:

$$\theta_{\mathrm{avg}} = \frac{M_{d\,\mathrm{avg}}}{K_r}$$

Here, $M_{d\,\mathrm{avg}}$ is the average value of the driving moment. Since this is proportional to the DC component of the input signal, the meter will only indicate the average value or DC component of the input quantity.

For a good step response, the relative damping z of the meter must be approximately equal to 0.7. Then, the overshoot after a step input will be only 4%. The bandwidth of the meter is f_0 and is equal to the frequency of the free undamped oscillations of the pointer of the meter. Generally, f_0 lies somewhere between 0.5 Hz and 10 Hz. The *response time* t_r, i.e. the time necessary for the indicated value to be within 1% of the final value, becomes $t_r \approx 1/f_0$, so t_r lies between 0.1 s and 2 s (see also Fig. 2.43).

3.4.2 Electro-optic displays

An electro-optic display is an output transducer which is used mostly for digital displays. These displays can be designed actively to emit light, or, to change their optical characteristics under the influence of an electrical signal. Then, they can be read-out under external illumination. The main advantage of the latter passive type of electro-optic displays is its very low power consumption (approximately 10 μW/cm^2, light emitting displays require at least 100 mW/cm^2). These passive displays are based on so-called *liquid crystals*. Certain organic substances combine the characteristics of both solids and liquids. Their molecules have the large degree of freedom of a fluid, but still retain mutual ordering, similar to the ions in the crystal lattice of a solid. This liquid crystal state, however, only exists over a certain temperature range. Below this, the material behaves as a solid and above this it behaves as a liquid. Within the usable temperature range, most molecules are aligned in the same, preferred direction. Many characteristics of liquid crystals are dependent on the orientation (anisotropic), such as the permittivity (ε_r), the refractive index and the transparency of the material. If an electrical field is applied, the orientation of the molecules and, therefore, the mentioned characteristics will be influenced. Fig. 3.72 depicts how this can be accomplished. The liquid crystal (thickness ≈ 10 μm) is enclosed between two glass plates, on the inside of which two extremely thin, transparent conducting metal (tin oxide) electrodes have been deposited. The surface of the glass is treated in such a way that the direction of the molecules is fixed with respect to the glass plate (director in Fig. 3.72(b)). Thus, the molecules on the two extreme opposite sides can be arranged at 90° to one another and the orientation of the molecules in between will gradually rotate. The incident light first passes through a polariser located in front of this arrangement. The polarisation of the light will gradually rotate as it travels through the molecules. The rotated light can now pass through an output polariser, which is placed at 90° with respect to the first. When a voltage is applied to the electrodes, the molecules will realign themselves along the electrical field (as in Fig. 3.72(c)). The polarisation of the incoming light is no longer rotated and will not pass through the cell; the cell appears dark. The required voltage is no larger than 1 or 2 V. When the polarisers are oriented in the same direction, the reverse effect occurs. If a mirror is placed behind the cell, the reflected light can be observed from the same side as the incident light. If a DC voltage is used to control the cell, galvanic effects may occur which will reduce the lifetime of the cell considerably. Therefore, low-frequency (50 Hz) AC voltages are usually used to drive the cell. The most commonly used liquid crystal uses nematic fluids (Gr. *nematos* = thread). Liquid Crystal Displays (LCDs) based on this are often referred to as 'twisted nematic' LCDs. For applications which must also be read in the dark it is possible to use the LCDs shown in Figs. 3.72(b) and (c) with back lighting. Unfortunately, though, most of the advantage of the low power consumption is then lost.

Due to the simple structure of an LCD cell, it can conveniently be given the shape of numbers, letters and other symbols. LCDs are relatively slow (turn-on time 50 ms, turn-off time 100 ms). The lower the temperature, the slower the LCD.

Light emitting electro-optic displays are based either on *incandescence* or *luminescence*. Incandescent displays make use of a filament, which converts electrical energy predominantly into heat and also some visible light. The power efficiency of this type of display is rather low. They are sometimes used in segment displays. A single filament requires roughly 3–5 V and 20 mA. Therefore, the power consumption per segment is somewhere between 60 to 100 mW.

Luminescent displays are based on the effect that when excited electrons return to their original energy state, they will emit visible light. The electrons of a suitable material can be excited in several ways, for instance, by the electrical discharge in a gas, by recombination or by bombardment with high-energy electrons.

Gas discharge displays consist of a glass tube filled with a gas and several wire electrodes in the shape of the required symbols, which are placed closely behind one another. These electrodes are the cathodes of the gas

Figure 3.72. (a) Structure of an LCD cell. (b) Rotation of the orientation of the molecules (and thus of the polarisation of the incident light) when no electrical field is applied. (c) Alignment of the molecules by applying an electrical field (no polarisation rotation).

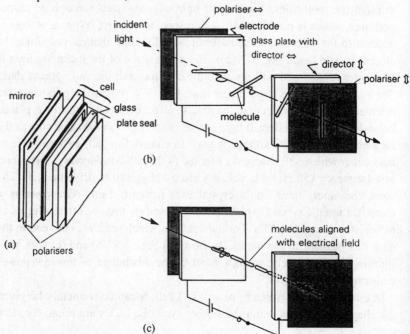

tube, which also contains a single anode. The pressure of the gas, the distances between the electrodes and the current density are chosen such that the cathodes are neatly covered by a layer of luminescent gas (cathode glow). These tubes are used almost exclusively for alpha-numerical displays and are called neon tubes or sometimes by their trade name, nixie tubes. The starting voltage is about 120 V, the operating voltage 80 V and the operating current 1 mA. The use of these tubes has declined dramatically with the advent of light emitting diodes.

In Light Emitting Diodes (LEDs), the electrons of the atoms are excited by means of recombination. A GaAs or GaP semiconductor diode will emit light when forward biased. The colour of the light can be changed by doping the semiconductor with various substances. This type of diode is often used in matrix displays. A common alpha-numerical display consists of 5×7 diodes. The supply voltage lies around 1.6–3 V and the supply current must be approximately 150 mA for sufficient brightness.

Electro-optic displays which employ luminescence by bombarding a substance with fast electrons are called cathode-ray tubes: an historic name, since in the past it was believed that there was such a thing as cathode rays. Electron beam tubes would be a more appropriate name. Besides numbers, these tubes can also display diagrams, curves and symbols. As this type of display is far more versatile than the other types of displays, we will discuss the cathode-ray tube more extensively.

Cathode-ray tube
Fig. 3.73 shows that four processes take place in a Cathode-Ray Tube (CRT): the *generation*, the *focusing* and the *deflection* of an electron beam

Figure 3.73. The structure of the various electrodes in a cathode-ray tube.

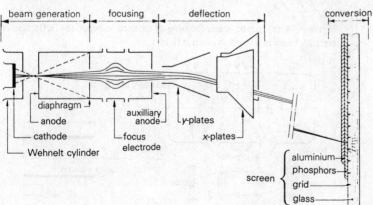

and the *conversion* of electrical energy into visible light. The electron beam is generated by a triode arrangement, which consists of a cathode, a control grid and an anode. The front of a cylindrical cathode is coated with a material (a mixture of barium and strontium oxide) with a low emission potential. By heating the cathode with a filament, the fastest electrons are emitted from the cathode. These are then caught in an electrical field and accelerated towards the anode along the central axis of the tube. The control grid between the cathode and anode is maintained at a negative voltage and is used to regulate the number of electrons reaching the anode (the beam current). When this grid is made more negative, the electron beam will also become narrower. The figure shows that electrons departing from the cathode at different places all travel through a single point. In reality, this is not exactly a point, but a small area, due to the repelling forces between electrons close together.

The beam focusing will project this small area as a dot on the screen of the CRT. This process has a lot in common with focusing by optical lenses and, therefore, one often speaks of electron lenses (see Fig. 3.74(a)). The focusing electrode between the two anodes is held at a lower potential than the anodes, which produces the electrical field indicated in Fig. 3.74(a). The electron beam entering the lens on the left will first encounter convex equipotential surfaces of a decreasing potential, which exert a dispersive force on the electrons. This can be compared to a diverging lens. The beam then travels through a decreasing concave field, in which converging forces are exerted on the electrons. This, in turn, is followed by another convex field, but now with an increasing potential. This field also converges the electron beam. Finally, the beam passes through a concave field of increasing potential, which diverges the beam again. One would expect, for a symmetrical electrode structure, a net result of zero; i.e. a parallel beam

Figure 3.74. (a) Electro-static focusing of an electron beam. (b) Deflection of an electron beam in a uniform electric field.

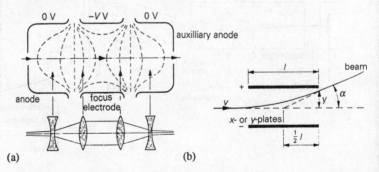

(a)   (b)

would leave the lenses again as a parallel beam. However, this is not the case, as the electrons are decelerated while they travel through the focusing electrode, due to its negative potential with respect to the two anodes. Therefore, the electrons remain for a longer period in the converging field and the net result is a converging system. This also holds true when the potential of the focusing electrode is positive with regard to the anodes. Electronic lenses always converge. Therefore, the potential U of the focusing electrode can be used to adjust the sharpness of the dot on the screen. If the density of the electron beam is altered with the control grid potential, the beam focus will change slightly and the dot on the screen will become a little bigger. When the intensity is changed (and therewith the screen intensity), the focusing must be readjusted.

The electron beam can be deflected magnetically as well as electrically. However, we will only discuss electrical deflection, as this is almost always used for measurement purposes since the beam must be deflected at very high rates to make high-frequency signals visible. Magnetic deflection is not suitable for this. The deflection electrodes (plates) are usually bent and tapered, as shown in Fig. 3.73. This is to allow a larger angle of beam deflection.

Let us now determine the deflection sensitivity of a CRT. Assume that the deflection plates are perfectly flat and parallel and that the electrical field between the plates of the capacitor thus formed is uniform. If the distance between the plates is d and the voltage across the plates is V, the field strength is given by $E = V/d$. This will exert a Coulomb force $F_e = qE = ma$ on an electron in this field (charge q, mass m, and acceleration a). As a result, the electron will be accelerated sideways at an acceleration $a = qV/md$. It will therefore follow a parabolic path. The vertical displacement y away from the axis of the deflection system is thus $y = \frac{1}{2} at^2$. The axial velocity v, with which the electron enters the capacitor, will remain constant and therefore the distance y in Fig. 3.74(b) can be expressed as $y = qVl^2/2mdv^2$. The angle α then follows from the expression:

$$\tan \alpha = \frac{dy}{dl} = \frac{qVl}{mdv^2}$$

It therefore appears as if the path of the electron is abruptly deflected by an angle α, exactly in the centre of the capacitor. The deflection sensitivity is large when the plates are long and close together and when the velocity of the electrons is low. Denoting the cathode–anode voltage of the electron gun by V_{ca}, we can calculate the velocity v of the electrons close to the screen using $\frac{1}{2} mv^2 = qV_{ca}$. Thus, tan $\alpha = Vl/2dV_{ca}$. The deflection sensitivity therefore decreases with an increasing anode potential.

If we wish to display very high-frequency signals (frequencies in excess of 10 MHz), the electron beam will travel across the screen extremely rapidly and the intensity of the emitted light will be low. This can be solved partially by increasing the beam current, though this will shorten the useful life of the cathode. Therefore, in high-frequency tubes the electrons are often preaccelerated to several kilovolts, then deflected and only then accelerated again (to approximately 15 kV).

At high frequencies the capacitance of the deflection plates nearly short-circuits the output of the *x* and *y* deflection amplifier. To solve this problem several plates are often used, which together form a transmission line with a characteristic resistance $R = \sqrt{\Delta L/\Delta C}$ (see Fig. 3.75). This way it is possible to realise extremely high-speed deflection systems of up to 2 cm/ns. This method is so successful because the deflection signal 'travels' along the tube axis in phase with the beam electrons.

The screen of the CRT is coated with a layer of phosphors (ZnS with Cu, Al and other elements) the atoms of which are easily excited when struck by the electron beam. Upon relaxation they produce light (Latin *phosphorus* = light bearer). These substances are not related to the chemical element phosphor. A phosphor can exhibit two mechanisms for emitting light. During the bombardment with electrons, the phosphor will very quickly reach a certain level of emission (see Fig. 3.76). This is called *fluorescence*. After the excitation has stopped, the phosphor layer will still glow for some time. This after-glow phase is referred to as *phosphorescence*. The colour of the emitted light depends on the chemical composition of the layer. The colour of the phosphorescent phase usually differs from that of the

Figure 3.75. High-frequency deflection system for a CRT.

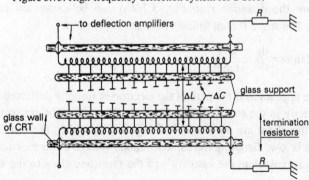

fluorescent phase. The duration of the phosphorescent phase also depends on the chemical composition. For visualising very slow phenomena (such as radar images), phosphors with a very long after-glow are used.

Often, the layer of phosphor is coated with a thin layer of aluminium on the side of the electron beam. This layer serves as a light reflector and also as a heat conductor. When the phosphor layer is bombarded with electrons, a large amount of heat is produced locally. The aluminium layer protects the screen from 'burn in' when the beam intensity is high and the deflection speed low. However, it does cost a certain amount of energy for the electrons to penetrate the aluminium layer. This can be offset by slightly raising the acceleration voltage of the CRT.

Nowadays, the grid of a CRT is usually etched directly into the back of the face plate. This eliminates the parallax caused by the thick safety glass of the tube.

Ambient light can easily produce disturbing reflections on the screen of a CRT and reduce the contrast. If the intensity of the electron beam is increased, the beam also defocuses and the resolution deteriorates. This problem can be avoided by placing an optical filter in front of the screen. The light from the screen is reduced only slightly, as it passes only once through the screen and the transmittance spectrum of the filter is chosen to fit the emission spectrum of the CRT. Ambient light, however, must pass through the filter twice and has a pretty wide spectrum. In this manner, reflections are absorbed considerably. The front of the filter is usually frosted to prevent reflections. Sometimes one also uses a fine carbon-black wire mesh to suppress ambient reflections.

Figure 3.76. Fluorescence and phosphorescence of a phosphor used in CRTs.

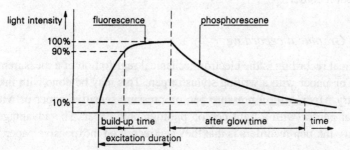

3.5 Recording

In Section 2.4 we described the process of recording measurement data and its purpose in measurement and instrumentation. The most important function of recording is to serve as a *memory*; we wish to save the information produced by the measurement for some time. In the first part of Chapter 2, we explained the difference between active and passive information. *Active information* is modulated onto some energetic carrier. Since there will always be a certain amount of energy dissipation, eventually, active information will be lost in the ever present thermal noise. If the energy content of the signal is increased by amplification, the noise will also be increased. Amplification of the signal at an earlier stage before it drowns in the noise will therefore only postpone the moment at which the signal is lost in the thermal noise. Therefore, information storage not be stored in the active form. *Passive information* is grafted into a particular ordered arrangement of matter. The information is as it were 'frozen' into the material. This type of information deteriorates far less quickly than active information. Also, we do not have to supply energy constantly to maintain storage. Therefore, we always convert active information into passive information before we can record it. This greatly limits the forms of storage available for data recording.

We can differentiate between *analogue* and *digital recording* in the same manner as we did earlier. In this text we will not deal with digital data recording, such as magnetic and magneto-optical memories used in tape drives, rewritable magnetic disk drives, etc.

Analogue electrical measurement signals are usually recorded in one of the following ways: graphically, magnetically or electro-optically. *Graphical recording* (Gr. *graphein* = to write) is the registration of a signal on paper with a pen. With *magnetic recording* the measurement signals are recorded on a magnetic storage medium. *Electronic recording* involves the recording of a measurement signal in the semiconductor memory of, for instance, a transient recorder.

3.5.1 Graphical cecording

Graphical recording is the electro-mechanical registration of a measurement signal on paper with a writing stylus or pen. This may be done with ink on ordinary paper, with a pressure pen on pressure sensitive paper or with a thermal pen or with a light pen on photographic paper. The advantage of ordinary ink pen recorders is that they can record on inexpensive paper. All other recorders need specially prepared paper. A disadvantage, though, is the friction between the pen and the paper which can give rise to recording errors. For this reason, high-speed recorders sometimes use ink-jets instead of ordinary pens. Pen recorders are less suitable for applications which require operation over a long period of time without supervision, as the pen can easily get clogged or run out of ink. For this kind of application other methods are preferred.

The simplest pen recorder is the *moving-coil pen recorder*. Fig. 3.77 shows that this recorder is based on a moving-coil meter with an arm and a pen, instead of a pointer. The heavy duty coil is, in fact, an electro-motor which moves the pen across the paper. The coil is usually driven by an amplifier. Thus, a high input impedance, a high sensitivity and a high bandwidth are achieved.

For a maximum pen swing of 5 cm, these pen recorders can achieve a bandwidth of approximately 100 Hz and a non-linearity of less than 1%. Due to their compactness, often, a number of moving-coil pen recorders are mounted next to each other. In this way, several signals can be recorded simultaneously on the same paper. This is especially useful when the relation in time between the recorded signals is important, as is the case, for instance, with electro-encephalograms and electro-cardiograms.

The trajectory or trace of the pen in this type of recorder is curved, since the pen will describe a part of a circle centred at the rotation axis of the coil. The curvature will result in distortion of the amplitude and time scale of the signal. The amplitude x of the recorded signal is then no longer proportional to the input voltage. We can determine the magnitude of the *amplitude distortion* as follows: If the angle of deflection of the arm is θ, the distance x in Fig. 3.77(a) is $x = r \sin \theta$, or with a series expansion:

$$x = r \left(\theta - \frac{\theta^3}{3!} + \frac{\theta^5}{5!} - \ldots \right)$$

Figure 3.77. Moving-coil pen recorder: (a) trace curvature; (b) arm mechanism to eliminate trace curvature.

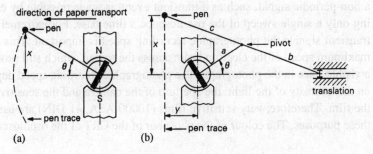

When the angle of deflection θ is small, this reduces to:

$$x \approx r\,\theta\left(1 - \frac{\theta^2}{6}\right)$$

The relative error in x is proportional to θ; at an angle of $\theta = \pm\frac{1}{4}$ rad ($\pm$ 14°) it will already amount to -1%. If the error is to remain small, the angle of deflection must remain small. The length of the arm may not be extended too far, as the friction between the pen and paper would produce too large a moment reducing the swing of the pen.

Besides amplitude distortion, the curvature of the trajectory of the pen will also cause *time scale distortion*. This can easily be seen by recording a square wave input signal. However, this effect can be eliminated almost entirely by using an arm mechanism such as in Fig. 3.77(b). If θ is so small that $\cos\theta$ can be approximated accurately by only the first two terms of a series expansion, then y is constant and independent of x if $b^2 = ac$. The apparent arm length of this construction is $r' = a(b + c)/b$. A necessary requirement, even when a special arm mechanism is used to reduce circular distortion, is that the angle of deflection θ remains small. This is because if θ is allowed to become large the time scale distortion will be much larger than in the case without an arm mechanism.

When considering photographic recording, we must distinguish between *direct* and *indirect* recording. The recorded signal in *direct recording* is (almost) instantaneously visible, without first developing the photographic paper. This is possible with paper which is made sensitive only to ultraviolet light. *Indirect recording*, on the other hand, requires the development of the photographic paper in the conventional way before a visible signal is available. The indirect method is used for recording (photographing) an oscilloscope image. Special cameras (oscilloscope cameras) are used for this purpose. In order to preserve the information on the amplitude and time scale of the signal, the grid on the face plate of the screen must also be photographed. The result can be made available immediately by using a transfer film (containing developer and fixative). Photographically recording a non-periodic signal, such as a transient event, is made possible by exposing only a single sweep of the oscilloscope's time base. For extremely fast transient signals the photographic recording speed is important. This is the maximum speed of the electron beam across the screen, which still produces a visible trace in the photograph. The photographic recording speed depends on the intensity of the light, the aperture of the camera and the sensitivity of the film. Therefore, very sensitive films (10000 ASA, 41 DIN) are used for these purposes. The colour of the phosphor of the CRT of the oscilloscope is also important. The human eye is most sensitive to greenish-yellow, so this colour would make perception of the signal of the oscilloscope easiest. However, the emulsion of a film is usually most sensitive to a bluish violet colour, so this is more suitable for producing good photographic recordings of the display. In addition, when photographing the display of an oscilloscope, better results can be obtained with prefogging, i.e. weakly illuminating the entire screen of the oscilloscope. This makes use of the fact that the differential sensitivity of the photographic emulsion depends on the intensity of the light. When the intensity is low, the differential sensitivity is lower than when the intensity is larger. If all of these precautions are observed, very high beam speeds of up to 3 cm/ns (30 000 km/s!) can be recorded.

The above mentioned methods of recording are all based on the deflection method. Very accurate recorders can be realised using the compensation method. Compensation of the input quantity in this kind of recorder is performed automatically using feedback. Such automatically compensating recorders are also called servo recorders. The principle of a servo recorder is illustrated in Fig. 3.78. The operation of an automatic compensator has already been discussed extensively in Section 3.3.2, so a few remarks will suffice here. The error signal $\Delta V = V_i - V_c$ is supplied to the input of a DC amplifier A_0 which has a large gain (to produce the large loop gain necessary to minimise the error signal). The null-offset errors of the amplifier are not suppressed by the feedback loop (additive disturbance). Therefore, an indirect DC amplifier is usually used. The amplifier drives a motor, which, in turn, moves a pen coupled to a resistive displacement sensor. The compensation voltage V_c is derived from a reference V_{ref}. If the closed-loop system is stable, the position of the pen will be adjusted so that $V_c \approx V_i$. From there on the pen position will follow the input voltage V_i. Thus, when the potentiometer is linear, the position of the pen will be a linear map of the

Figure 3.78. The principle of a servo pen recorder.

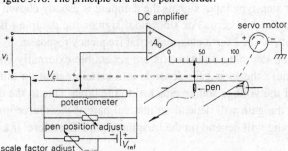

input voltage. The null position of the pen on the paper and the scale factor can be adjusted as indicated in Fig. 3.78.

Servo pen recorders have found numerous applications. They have all the advantages of the compensation method. The *input impedance* of a servo recorder is extremely high (once compensation is reached). The accuracy is determined by the accuracy of V_{ref} and the magnitude of the loop gain. The *linearity* depends on the linearity of the potentiometer used as a pen position transducer. This potentiometer should not therefore be electrically loaded. Another important characteristic of a servo recorder is the *reproducibility* of a recording. This is limited by amplifier offset drift, noise, hysteresis and a dead zone. In order to minimise the offset and drift, an indirect DC amplifier is used. The noise of the system should be small enough to stay within the line thickness of the pen. Hysteresis is caused by mechanical play in the connection of the displacement transducer to the pen and by the flexibility of the pen tip. Finally, a dead zone results from friction of the moving parts of the system, which gives rise to a threshold voltage, below which the servo-motor will not move.

The servo recorder of Fig. 3.78 will exhibit a second order dynamic response, due to the inertia of the moving parts (see Section 2.3.3.2). As the loop gain is increased, the relative damping z of the feedback system is reduced; the system becomes less damped; the frequency response will have a sharper peak and the step response will ring more and more. This behaviour is undesirable for a recorder which should record signals as faithfully as possible. To obtain a damping z which is as close as possible to critical damping, the velocity of the pen is fed back as well as its position. This is performed by subtracting a voltage from V_i, which is proportional to the speed of the servo motor. The magnitude of the proportionality deter-mines the resulting velocity damping. A recorder will therefore have two adjustments: one for the loop gain (*gain*) and one for the velocity feedback (*damping*). The gain should be adjusted high enough still just to allow the desired dynamic behaviour of the recorder to be set by adjusting the damping. For step-type input signals the damping is usually chosen so that $z = 1$. The overshoot is zero. For sinusoidal signals the damping is usually chosen so that $z = 1/\sqrt{2}$ for a maximally flat frequency response.

The gain and damping adjustments are accessible externally on a servo recorder because the loop gain depends to a certain extent on the internal impedance of the voltage source attached to the input. Due to the design of the recorder, the gain will depend on the real part of the source impedance and the damping will depend on the imaginary part. Therefore, if a recorder is to be used with very different signal sources, the gain and damping must be re-adjusted.

A servo pen recorder also exhibits *dynamic non-linearity*. One of the causes of this frequency dependent form of non-linearity is *velocity limitation*. The velocity of the pen is bound to a maximum which depends on the type of servo motor used. When the input signal varies quickly, the pen will no longer be capable of following it. This will result in an error which depends on dv/dt of the input signal $v(t)$. If the input signal is sinusoidal, i.e. if $v(t) = \hat{v} \sin \omega t$, and the non-distorted output signal is $y(t) = b \sin \omega t$, the velocity of the pen is largest at the zero crossings. Expressing $y(t)$ in centimetres and t in seconds, we can write this velocity as:

$$\left|\frac{dy}{dt}\right|_{max} = b\omega \ [\text{cm/s}]$$

The sine wave will therefore distort when $b\omega > v_{max}$. Here v_{max} is the maximum velocity (referred to as the slewing rate) of the servo system. In the plot of Fig. 3.79 the continuous curve marks this limit as a function of the amplitude and frequency of a sinusoidal output signal. Above the curve the recorder will become dynamically non-linear. Due to this velocity distortion, a sine wave will be distorted into a triangular wave (see Fig. 2.34).

Besides a maximum velocity, a servo system will also exhibit a maximum acceleration a_{max}. This is the largest acceleration that the pen can still follow. For a sinusoidal movement $y(t) = b \sin \omega t$ [cm], the acceleration is largest at the peaks of the curve. This acceleration is given by:

$$\left|\frac{d^2y}{dt^2}\right|_{max} = b\omega^2 \ [\text{cm/s}^2]$$

Figure 3.79. Dynamically non-linear region for a servo pen recorder writing sinusoidal signals: $b \sin 2\pi f t$.

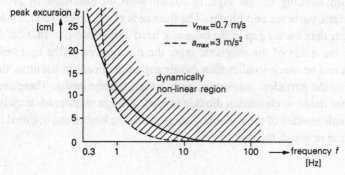

When $b\omega^2 > a_{max}$ *acceleration distortion* will occur, resulting in a rounding off of the peaks of the plotted curve. The maximum acceleration is determined by the inertia of the total moving mechanical system (servo motor, gears, driving belts, pen and runner) and the peak power that the DC amplifier can deliver into the servo motor. In Fig. 3.79 the dashed line indicates for which sinusoidal signals dynamic non-linearity occurs due to the maximum possible acceleration of the servo system. This figure shows that for low-frequency signals, the non-linearity results from the slewing rate-limitation, while for signals of small amplitude and high frequency, non-linearity is caused by the maximum acceleration of the servo system.

If the paper of a pen recorder travels past the pen at a constant rate, the recorder is referred to as an (x,t) recorder. The input signal is recorded as a function of time. It is also possible to move the pen with *two* automatic compensators, one in the direction of the x axis and one in the direction of the y axis. The paper remains stationary. In this case, we speak of a (x,y) recorder or *plotter*. An example of the specifications of a plotter is: full-scale width 25 cm; most sensitive range 50 μV/cm; static non-linearity less than 10^{-3} of full scale; reproducibility better than 10^{-3}; response time within $\pm 5\%$: 0.1 s; $a_{max} \approx 3$ m/s^2; $v_{max} \approx 0.7$ m/s.

3.5.2 Magnetic recording

A popular method of recording measurement results is by means of magnetic recording. This may be done in either analogue or digital form on a magnetic tape. We will limit our discussion to analogue magnetic recording, for which an instrumentation recorder is used (see Fig. 3.80) which can usually record many tracks (up to 14) adjacent one another on a tape.

The magnetic tape consists of a synthetic film (PVC or mylar) flexible enough to fit snugly the convex magnetic heads of the recorder, but strong enough not to stretch or expand, since this would distort the recorded time base.

This film backing of the tape is coated with an emulsion of small ferromagnetic particles and binder. The thickness of this layer need not be much larger than the air gap in the magnetic head of the recorder. In order to minimise the noise of the magnetic tape, the dimensions of the magnetic particles must be much smaller than the air gap of the head. In addition, the direction of the particles' magnetisation must not change easily. Therefore, usually iron oxide or chromium dioxide is used. A large number of densely-packed small needles of the material used is put in the binder and oriented in the direction in which the tape travels.

The signals are recorded by writing a pattern of magnetisation changes into the oxide layer. This pattern is done by the *write head*. Its principle is shown in Fig. 3.81. The inductive write head consists of a coil wound around a yoke with an air gap. The yoke is manufactured from a material with a high permeability, low hysteresis and minimal losses due to eddy currents, for instance, Sendust, or laminated mu-metal or permalloy. The magnetic field strength in front of the air gap is proportional to the current through the coil. As the strength of the magnetic field increases, the number of particles in the magnetic tape which are permanently magnetised in the same direction also increases. Unfortunately though, the relation between the strength of the magnetising field and the resulting magnetisation is very non-linear. This is evident with the aid of Fig. 3.82. In Fig. 3.82(a) the *BH* hysteresis loop of the ferromagnetic material of the tape has been plotted. The recording field strength H is proportional to the current flowing through

Figure 3.80. Analogue instrumentation recorder with direct recording and FM recording.

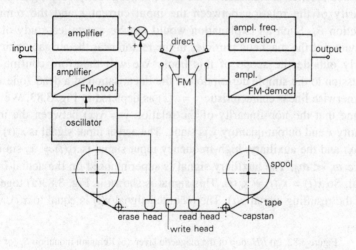

Figure 3.81. The inductive write head as an electromagnetic transducer.

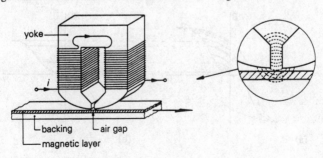

the head. Assuming this current is kept constant, imagine that a section of a previously demagnetised tape is passed underneath the head. The field strength which the section of tape experiences first rises to a maximum right under the air gap and then falls off again. The induction B will then follow the BH loop as indicated. So, when the field strength is zero again, a remnant induction B_r will remain. As the current increases, so does B_r. The relation between i and B_r and, hence the relation between H and B_r, is far from linear, as can be seen in Fig. 3.82(b). The relation between the magnetisation B_r and the current I has a dead zone, which will give rise to odd harmonic distortion of the recorded signal.

To improve the linearity of the recording, a high-frequency AC field is superimposed on the recording magnetic field. The frequency of this additional field is much greater (approximately five times) than that of the highest-frequency signal component.

We will demonstrate how the introduction of such a high-frequency signal can improve the linearity of a system. First though a remark concerning the exact and detailed explanation of the effect of a high-frequency signal on the linearity of the relation between the input current i and the remnant induction B_r. Such an explanation would require a detailed study of the behaviour of the magnetic particles in the region near the air gap, which is clearly outside the scope of this book. We will therefore confine our discussion to the simple description of the linearisation of a dead zone x_0 in an otherwise linear characteristic $y = y(x)$ as depicted in Fig. 3.83. We will assume that the non-linearity of the relation $y = y(x)$ between the input quantity x and output quantity y is static. The actual input signal is $x_s(t) = \hat{x}_s \sin \omega_s t$ and the auxiliary, high-frequency input signal is $x_a(t) = \hat{x}_a \sin \omega_a t$, where $\omega_s \ll \omega_a$. The auxiliary signal is superimposed on the actual input signal. So $x(t) = x_s(t) + x_a(t)$. This signal is shown in Fig. 3.83(a) together with the resulting signal $y(t)$. The recorded signal $y(t)$ is equal to $x(t)$, with

Figure 3.82. (a) BH loop of the magnetic layer. (b) Remnant induction B_r (or magnetisation) as a function of the field strength H.

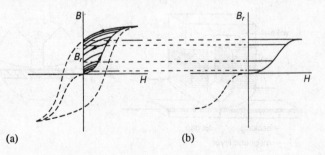

(a)   (b)

the exception of a band of width x_0 which is missing around the zero crossings. If the high-frequency components are filtered from $y(t)$ by a low-pass filter (with a cut-off frequency at, for instance, $\omega = \sqrt{\omega_s \omega_a}$), only the signal $y_s(t)$ will remain. This signal is almost perfectly sinusoidal. The distortion of $y_s(t)$ has been plotted in Fig. 3.83(b). When there is no auxiliary signal ($\hat{x}_a = 0$), the distortion is large. If the width of the dead band is $x_0 = 1$ V, input signals with $\hat{x}_s < 0.5$ V are lost completely. A small auxiliary signal can reduce the distortion considerably. For a certain value of $\hat{x}_s$, the distortion decreases when $\hat{x}_a$ increases, whilst for a given $\hat{x}_a$, the distortion decreases as $\hat{x}_s$ decreases. If $(\hat{x}_a - \hat{x}_s) < x_0/2$, the distortion will rise sharply, since the envelope of the input signal lies entirely within the dead band.

The ideal waveform for the auxiliary signal is a square wave. The only requirement for a square wave is $(\hat{x}_a - \hat{x}_s) > x_0/2$. If this condition is satisfied, the distortion is zero. However, perfectly square auxiliary signals are not practicable and, in addition, the non-linearity is not entirely static, so a certain non-zero level of distortion will always remain.

This method of linearisation is frequently applied. Examples are: tapping a (poor) meter whose pointer is stuck as a result of stiction; vibrating a

Figure 3.83. Linearisation of a dead band using an auxiliary signal. (a) The input signal $x_s(t)$ on which the auxiliary signal $x_a(t)$ is superimposed. The width of the dead zone is x_0. After low-pass filtering, the output signal is $y_s(t)$. (b) The distortion D of the output signal $y_s(t)$ as a function of the peak amplitude $\hat{x}_s$ of the input signal, with a dead zone of 1 V and various values of the peak amplitude $\hat{x}_a$ of the auxiliary signal.

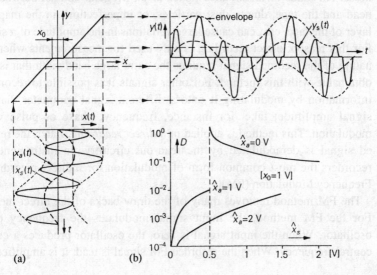

mechanical transmission which has play; addition of a dither signal to linearise ADCs, etc.

If the magnetic tape travels with velocity v, a periodic write signal of frequency f will result in a magnetic pattern on the tape with wavelength $\lambda = v/f$. When the wavelength λ approaches the width w of the air gap, the write process collapses, since the region that was just written is rewritten by the following write current change. Therefore, $\lambda > w$ must hold. The bandwidth of the write head increases with a narrower air gap and a higher velocity v.

The magnetic pattern on the tape is retrieved with a *read head*. This head has the reverse function of the write head. Since these electromagnetic transducers are reversible, both heads are based on the same principle. The passing, differently magnetised regions will cause a change of flux in the head yoke and induce a voltage $v = -n \, d\Phi/dt$ in the coil of the head. Since $\Phi = c_1 B_r$ and $B_r = c_2 H$ and $H = c_3 i$, the readback voltage v is related to the current i in the coil according to $v = c \, di/dt$. Obviously, magnetic recording has a differentiating effect on the signal. The gain of the transfer function increases by a factor of 2, if the frequency of the signal is doubled. The read-out amplifier must therefore perform a frequency correction in the form of an integration. It is also evident that this method of recording is not capable of recording DC signals, as these signals would produce a constant magnetisation and a constant flux. Consequently, no voltage would be induced in the read head.

The method of recording discussed above is sometimes referred to as *direct recording*. This method gives rise to a high level of multiplicative noise. All kinds of disturbances, such as an irregular contact between the head and the tape, due to dust particles, or imperfections in the magnetic layer of the tape, etc., can cause large variations in the amplitude of a signal. For this reason, direct recording is only used for measurements where the nature of the recorded signal requires the extra large bandwidth that is only obtainable with this method. For other signals it is possible to record the information by modulating it onto a different signal parameter than the signal amplitude; take, for instance, frequency, phase or pulse width modulation. This method is applied in *indirect recording*. Later, the retrieved signal is demodulated by the read-out circuitry. In instrumentation recorders the most common form of modulation for indirect recording is Frequency Modulation (FM).

The FM method removes many of the draw-backs of the direct method. For the FM method, the input signal modulates the frequency of an oscillator. When the input signal is zero, the oscillator produces a certain central frequency. When the recorded FM signal is read, it is amplified and limited (as the amplitude does not contain the recorded information) and converted back into a voltage, which is proportional to the deviation between the central and the actual frequencies, by an FM demodulator. The great advantage of the FM method is that the imperfections and noise of the magnetic tape do not cause any problems, so the amplitude of the demodulated signal is more stable. Furthermore, it is no longer necessary for the magnetisation to be linearly proportional to the recording field. Therefore, the linearity of the demodulated FM signal is also higher. Finally, another advantage is that the FM method is capable of recording DC signals. However, the price that must be paid is a reduction of the bandwidth by about a factor of 10. The FM method also offers the possibility of correcting for variations of the tape speed.

The specifications of instrumentation recorders have been divided into four classes, which are labelled: 'low band', 'intermediate band', 'wide band, group 1' and 'wide band, group 2'. The tape speed has been standardised: 304.8 cm/s, 152.4 cm/s, 76.2 cm/s, ... 4.7625 cm/s. Each following speed is half that of the preceding value. Consequently, the bandwidth is also halved for each step.

The four classes are arranged according to increasing bandwidth and decreasing signal-to-noise ratio. A direct recording with a tape speed of 304.8 cm/s and a recorder from the 'intermediate band' class will provide a bandwidth from 300 Hz to 600 kHz and a signal-to-noise ratio of 40 dB. For a 'wide band, group 2' recorder these figures are 500 Hz to 2 MHz and 22 dB.

This also applies to FM recording. A 'low band' recorder with a tape speed of 304.8 cm/s, a central frequency of 108 kHz and a frequency deviation of ± 40% will provide a –1 dB bandwidth of 0–20 kHz and a signal-to-noise ratio of 50 dB. A recorder from the 'wide band, group 2' class with a central frequency of 900 kHz and a deviation of ± 30% will have a –3 dB point at 400 kHz and a signal-to-noise ratio of 36 dB.

An advantage of magnetic recording is that the time base can easily be compressed or expanded by simply varying the tape speed. In this way, it is possible to shift high- or low-frequency signals within the frequency range of the measurement equipment. It is also possible to record many tracks on a single tape, thus not only recording the variations of the signals in time, but also the information on their coherence. This is accomplished by using recording heads which are constructed with a row of head and write heads. A recorded signal may be converted into an electrical signal time and time again, which is useful for analysis. Furthermore, a tape may be erased, cut and edited. Finally, the information density of a magnetic tape can be extremely high (i.e. it stores very large amounts of measurement data).

3.5.3   Electro-optical recording

The recording speed of the discussed methods of recording is very limited, because in every case moving mechanical parts are required. Often, though, one wants to record very high-speed phenomena for further examination and analysis. This is made possible by fast solid-state memories. If the input signal is sampled at a very high rate and stored in a semiconductor memory, it can be read out later at a much lower speed and displayed, for instance, on an oscilloscope or pen plotter. This results in an effective enlargement of the time scale. In this manner, time scale expansion factors of 10^9 (so that a nanosecond becomes a second) can easily be accomplished. It is as if the phenomena occur 10^9 times slower. Instruments which are capable of this are called *transient recorders*. It must be noted that though a sampling oscilloscope also samples the input signal (see Section 4.4), it can only display *periodic input signals*, as it is based on the principle of *coherent sampling* (see Section 2.2). This type of oscilloscope does not store the signals, but displays them directly on the screen.

Fig. 3.84 illustrates the principle of a transient recorder. This instrument is sometimes also referred to as a *digitising oscilloscope*, though this name does also include the above mentioned sampling oscilloscope. An essential function of a transient recorder is that it can record both periodic and aperiodic (transient) signals, which can then be displayed later, at a much lower rate, and in a periodic fashion.

An electronic transient recorder operates as follows: the input signal is first converted at equidistant moments in time into (digital) data by an ADC. These data are written into a cyclic semiconductor memory. This is a

Figure 3.84. Electronic recording by means of a transient recorder. The recording is performed by digitising the input signal, which is stored in a cyclic memory. The signal is played back as an analogue signal for later use.

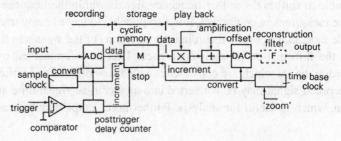

continuous process in the absence of an input trigger signal; new data over-writes old data continuously. The memory determines the amplitude resolution (word length) and the length of the trace (number of stored words). If an input trigger signal is given, a comparator activates a (posttrigger) counter. This counter, which is set to a certain initial value externally, determines the number of samples which will be stored in the semiconductor memory, before the cyclic storage process is stopped. For example, if the size of the memory is 4000 words, the trigger delay counter may be set to an initial value of 2000. Now, after the contents of this counter becomes zero, 2000 samples *before* and 2000 samples *after* the trigger point will be stored in the memory. Because the sample clock is running at predetermined accurately known frequencies, the reproduction of the signal contains information on both the amplitude and the shape of the signal versus time. If, for instance, the clock frequency is set to 500 Msamples/s, we can observe an interval of $\pm$ 4 μs around the trigger point; the samples are taken every 2 ns. The process of continuously writing over old data which is stopped at a point in time delayed with respect to the trigger point allows us to view the events which happened *before* the trigger occurred. After the transient recorder is stopped, the digital data are immediately available for signal processing by a computer. They can also be converted back to an analogue signal, as shown in Fig. 3.84. This reconstructed signal can be made available at any (adjustable) rate, which can be set by the time base clock generator. If, in the example given above, this rate is set to 5 pulses per second, 10 ns of the input signal will correspond to 1 second of the output signal (an expansion factor of 10^8). This signal could then be simply recorded by a pen plotter, which would take 13.3 minutes in total. If the time base clock generator scans only a fraction of the memory, we can 'zoom in' on the signal and only display its most interesting part. When the signal is converted back to an analogue voltage signal, the amplitude of the signal can very easily be altered (amplified or attenuated) by digital multiplication and an offset can be introduced (adding a binary number). In order to obtain an analogue reconstruction of the input signal which faithfully follows the input, the incremental steps in the output signal of the DAC must be removed by a low-pass filter. We will return to these so-called reconstruction filters in Section 4.5. For a true reconstruction, the (primary) sampling frequency must be at least twice that of the highest component in the input signal. This will also be dealt with in Section 4.5. With the sampling rate of 500 Msamples in the above example, the bandwidth of the input signal must be less than 250 MHz. However, to minimise errors (5%), in practice the bandwidth would have to be limited to 50 MHz.

It is apparent that the rate at which the *primary* samples are taken governs the frequency range of the transient recorder (the *secondary* samples determine the pace of which the time base clock generator takes the data out of the memory). The necessary time for each primary sample is determined by the sampling circuitry, the conversion process and the storage in memory. Therefore, several alternatives for simple, straightforward equidistant sampling have been devised (see Fig. 3.85). In Fig. 3.85(a) the input signal is sampled quickly (compared to the fluctuations of the signal) at equispaced intervals. This is the case that we discussed above. This is called 'real time' sampling and requires extremely fast sampling and ADC circuits.

Fig. 3.85(b) shows how the same signal can be sampled coherently (or sequentially). Now, the signal *must be periodic*! In the first period of the signal a sample is taken at point '1'. In the next period a sample is taken at '2'. (In fact, the sampler may skip an integer number of periods before it takes the next sample.) It will be clear that in this manner the conversion and storage may be performed at leisure; there is plenty of time. The difficulty lies in accurately maintaining the relation in time between the samples and the period of the input signal. As the time between samples increases, by skipping periods, it becomes more difficult for the samples to stay coherent with the measured signal. On the other hand, if fewer periods of the input signal are dropped between taking samples, the circuitry must again respond more rapidly.

Fig. 3.85(c) shows a hybrid form of sampling, called random repetitive sampling. The first signal period is sampled coarsely, at equispaced sampling points indicated as '1'. The second period (again, after possibly skipping a number of periods) is also sampled coarsely, at the same equispaced intervals. However, the samples '2' are placed totally randomly with respect to the first samples. Here, the samples no longer need to be coherent with the input signal. The only requirement is that the distance between samples is constant and that the signal is periodic. The input signal can now be reconstructed by simply measuring the time elapse between the trigger point and the sampling clock.

Figure 3.85. Various principles for sampling measurement signals: (a) equispaced, real time sampling; (b) coherent, sequential sampling; (c) random, repetitive sampling. NB The last two require the input signal to be periodic in nature.

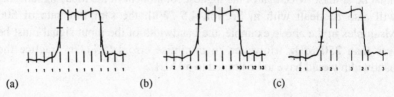

(a)   (b)   (c)

4

Electronic measurement systems

Certain combinations of signal processing and display are used so frequently in measurement systems that they have become instruments in their own right. This applies especially to equipment which is designed for measuring such fundamental quantities as frequency, amplitude (either current or voltage), resistance, etc. Equipment which is designed for frequently recurring signal analysis, such as the oscilloscope in the time domain or the spectrum analysers in the frequency domain also falls in this category. These quantities are measured virtually exclusively with electronic equipment. In this chapter we will examine a number of these electronic measurement systems.

4.1 Frequency measurement

The repetition frequency of a phenomenon is the number of times this phenomenon occurs per unit of time. The time interval between two repetitions is called the period T. Hence, frequency and period are closely related ($f = 1/T$). Therefore, a measurement of frequency can often be replaced by a time interval measurement and *vice versa*.

There are many ways of measuring the frequency of an electrical signal. One method is the *resonance method*, which was used in the past in a frequency meter containing a number of bars which were forced to vibrate by the input signal. Each bar was tuned to a slightly different resonance frequency. The frequency of the input signal was taken to be roughly equal to the resonance frequency of the bar with the largest amplitude. (This measurement method was used for determining the AC line frequency of auxiliary power generators.)

257

Another method of measuring frequency is based on *frequency-to-voltage conversion* (see Fig. 4.1). In this the input voltage V_i is first converted into a rectangular wave form. The rising edges of the rectangular wave trigger a monostable multivibrator, which produces an input-synchronised narrow pulse with a constant height V and a constant duration T. The average (or DC) value of these pulses is determined by a low-pass filter and subsequently displayed by a meter. The range of this type of voltage converter is approximately two decades. At high frequencies, the frequency range is limited by the rise time of the rectangular wave and the rise and fall time of the pulses. At low frequencies, it is limited by the low-pass filter. The cut-off frequency of this filter must be low enough to minimise the ripple in the output to the meter. This ripple results from the not completely filtered pulsed voltage V_B. To accommodate low-frequency measurements the cut-off point of the filter must be pushed down, which makes the response of the system very sluggish.

A *time-interval-to-voltage converter* does not have the latter disadvantage. This converter produces a voltage which is proportional to the time between two consecutive zero-crossings of the input signal. So, in fact, it measures the period as opposed to the frequency. The first section of the converter is identical to that of Fig. 4.1. The pulses V_B are compared to a sawtooth voltage (see Fig. 4.2). This sawtooth voltage V_S has a well-defined slope and is reset at the positive edges of the pulses V_B. At the time at which a pulse

Figure 4.1. (a) The principle of a frequency-to-voltage converter. (b) Wave shapes as a function of time in the converter (a).

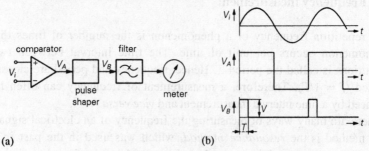

(a)   (b)

Figure 4.2. Wave shapes in a time-interval-to-voltage converter.

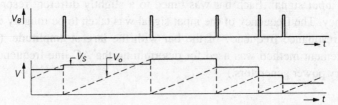

occurs, the voltage V_S has reached a value, which is stored by a hold circuit for precisely one period. The output voltage V_o thus obtained is a proportional measure for the duration of the previous period. For large periods, the measurement range is limited by the maximum amplitude of the sawtooth that can be realised. For small periods it is limited by the reset time of the sawtooth. The measurement range is also approximately two decades.

The most widely used method of frequency and period measurement is the *enumeration method* used in *frequency* and *period counters*. Strictly speaking, the name is incorrect, as only numbers can be counted and not quantities with a physical dimension such as frequencies or period times. In a frequency counter the number of oscillations of the measurement signal, which fall within a single period of an accurately defined reference frequency signal is counted. The procedure is simply reversed in a period counter, i.e. the number of periods of a reference signal which fit a single period of the measurement signal is counted. Therefore, the same counter can perform both measurements by interchanging the two signals. The great popularity of this method of measurement is mainly due to the rapid development of fast digital circuits and the availability of ultra-stable, accurate quartz-crystal controlled oscillators.

A frequency counter converts the input signal into a square wave of the same frequency with the aid of a comparator (see Fig. 4.3(a)). During the time the gate is open (conducting), the periods of the input voltage V_i are counted by a decimal counter. After the gate closes, the accumulated count is made visible by an alpha-numerical display. The length of the time interval during which the gate is open (the gate time) is determined by the frequency f_0 of a reference oscillator. This frequency f_0 is divided by means of an adjustable frequency divider by powers of 10 (decade counter). The gate time is now made equal to half a single period of this divided reference frequency. Therefore, the gate control frequency must still be divided by a factor of 2 between the output of the decade counter and the gate, assuming the gate is open when the output is low, and closed when the output is high. The gate time is then equal to $10^n/f_0$, where n is the setting of the decade counter (which divides by 10^n). Thus, we find for the number of counted periods of the input signal at the end of the gate time:

$$c = \frac{10^n f_i}{f_0} \pm 1$$

in which f_i denotes the frequency of the input signal. We see that the enumeration method is used to count a ratio; we are determining the magnitude of f_i with respect to f_0. This reference frequency f_0 must therefore be accurately defined.

The reference signal is produced by a quartz-crystal oscillator. A well-cut quartz crystal can be used as a mechanical resonator with a very high quality factor (10^6–10^7). The piezoelectric crystal is used to stabilise the frequency of an electronic oscillator. Often, the entire oscillator is placed in an oven which is controlled by a thermostat to maintain a constant temperature, even when the oscillator is not being used. In this way, a short-term stability of 5×10^{-11} per hour and a frequency drift — due to ageing — of less than 5×10^{-10} per day have been achieved.

The maximum frequency which can be measured in this manner is determined by the speed of the digital counters. Frequencies up to 150–200 MHz can still be measured in this way. The lower limit for the frequency is determined by the long gate time required for counting a reasonable number of periods. The accumulation of a reasonable number of periods is necessary to ensure a small *quantisation error*, which occurs with every gated count. As the instant at which the gate is switched is not related to the zero-crossings of the input signal, the quantisation error is equal to plus or minus one period of the input signal.

Especially at low frequencies, when the display and the counter are not entirely filled, the quantisation error will give rise to a relatively large error in the measured frequency.

Take, for instance, the following example: If $f_0 = 1$ MHz and the maximum frequency division factor is $n = 10^7$, the largest gate time that can be selected is 10 s. A frequency of precisely 900 kHz will then be displayed as $9\,000\,000 \pm 1$, whereas a frequency of 9 Hz is measured as 90 ± 1. Therefore, the quantisation error of a frequency counter varies proportionally to the reciprocal of the frequency. Evidently, it also varies inversely proportionally to the gate time.

For measuring low frequencies it is preferable to exchange the frequency measurement for a *period measurement*. This is easily accomplished by interchanging the input signal and the signal of the reference oscillator (see Fig. 4.3(b)). The gate time is now equal to the period $1/f_i$ of the input signal V_i. For the duration of this interval, the counter now counts the number of cycles of the divided-down frequency of the reference oscillator. Thus, the count at the end of the gate time is:

$$c = \frac{f_0}{10^n f_i} \pm 1$$

In contrast to the previous situation, the count accumulated in the counter will be large when f_i is small. Hence, as the frequency decreases, the influence of the quantisation error (the term ± 1 in the expression for c) becomes less. The relative quantisation error in a period measurement is proportional to the input frequency. (The opposite is the case with a frequency measurement, the error decreases with increasing frequency.) In a period measurement the relative quantisation error is inversely proportional to the divided reference frequency $f_0/10^n$.

Figure 4.3. Frequency and period measurement according to the enumeration method: (a) frequency counter; (b) period counter; (c) period counter using averaging.

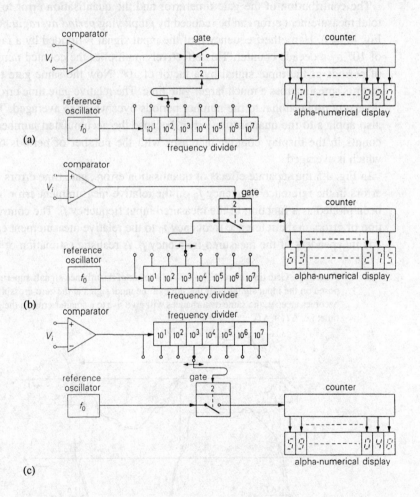

(a)

(b)

(c)

A second error which occurs in period measurements, in addition to the quantisation error, is the *gate time error*. As can be seen in Fig. 4.4, the input signal V_i contains noise, distortion and interference. The zero-crossings of the signal will therefore no longer be as well defined as in a clean signal. The gate time in the example shown in Fig. 4.4 is no longer T but is equal to $T + \Delta T_1 + \Delta T_2$. The relative error of the gate time is given by the ratio $(\Delta T_1 + \Delta T_2)/T$. As the slope dV_i/dt of the input signal V_i at the zero-crossings increases the gate time error will become smaller, if the disturbances remain the same. Clearly, the ideal situation is an input with a rectangular wave shape.

The contribution of the gate time error and the quantisation error to the total measurement error can be reduced by employing *period averaging* (see Fig. 4.3(c)). Here, the frequency f_i of the input signal is divided by a factor of 10^n by a decade counter. This effectively multiplies the counted number of periods of the input signal by a factor of 10^n. Now the same gate time error is spread across a much larger gate time. The relative gate time error is inversely proportional to the number periods over which is averaged. This also applies to the quantisation error, because the accumulated number of counts in the display counter increases with the number of periods over which is averaged.

In Fig. 4.5 the separate effects of quantisation errors, gate time errors and errors in the reference frequency f_0 on the relative measurement error have been plotted as a function of the measured input frequency f_i. The contribution of errors in the reference frequency f_0 to the relative measurement error ε is independent of the measured frequency f_i. A realistic estimation of this

Figure 4.4. Gate time error $\Delta T_1 + \Delta T_2$ as a result of disturbance signals superimposed on the input signal V_j. If the slope of the input signal at the zero-crossings becomes steeper, the same disturbances will give rise to a smaller error in the gate time, i.e. $\Delta T_1' + \Delta T_2'$.

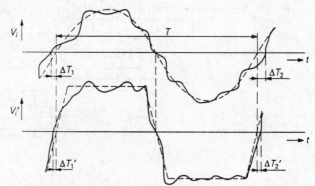

type of error is 3×10^{-7}. The contribution of the quantisation error is frequency dependent. For a frequency measurement the quantisation error is proportional to the reciprocal value of the measured frequency f_i. For a period measurement it is proportional to the frequency f_i. This gives the slopes in the error lines of Fig. 4.5. The quantisation error is inversely proportional to the gate time and also inversely proportional to the number of periods over which is averaged. This also produces a series of sloped error lines. The absolute position of these lines can be determined as follows. Assuming that the reference frequency $f_0 = 1$ MHz, then if the gate time is exactly 1 s, the value 1 000 000 will be displayed. The relative quantisation error then amounts to $\varepsilon = 10^{-6}$. This point is marked A in Fig. 4.5. If a period measurement is performed for only one period, with an input frequency of, for instance, 10 kHz, the displayed value is 100. Then the relative quantisation error is $\varepsilon = 10^{-2}$. This gives point B. At low frequencies, the error in a period measurement is determined by the gate time error. Assuming that the input signal's signal-to-noise ratio is independent of frequency, this results in a constant relative gate time error. A realistic value of this error is 3×10^{-4}. If the average over 10^n periods is taken, the gate time error will be a factor 10^n smaller. The gate time error causes the error curves to level off in the left-hand side of Fig. 4.5.

4.2 Phase meters

There are many ways of measuring the phase difference between two periodic signals. In power electronics, an electrodynamic meter consisting of

Figure 4.5. The relative measurement error ε for frequency counting and period counting due to quantisation errors, gate time errors and errors in the reference frequency f_0. In this example $f_0 = 1$ MHz.

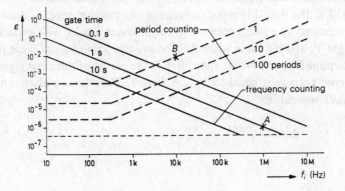

two perpendicular moving coils is often used for measuring the phase difference between an AC current and an AC voltage.

An oscilloscope can be used for roughly determining the phase difference between two signals. If one signal is connected to the vertical or *y*-input and the other to the horizontal or *x*-input an elliptical Lissajous figure is created from which the phase difference can be determined. A difference in phase shift of the two input amplifiers of the oscilloscope will result in a measurement error here. However, this phase difference can easily be measured by connecting the same signal to both inputs. This value is then subtracted from the measured phase difference.

The phase difference of two signals can also be determined with an oscilloscope by performing two time interval measurements. A dual-beam oscilloscope allows us to measure the time difference Δt between the zero-crossings of two signals. If the period T of the signals is known, we can calculate the phase difference as $\phi = 2\pi \, \Delta t/T$ radians. However, phase measurements with an oscilloscope are not very accurate. For a more accurate measurement, other methods are available.

In Fig. 4.6 the *phase compensation method* is illustrated. Block A represents an adjustable phase giving an accurately known phase shift. First, the same signal (V_1 or V_2) is connected to both inputs in order to set the output of the system to zero by adjusting B. Then, the two signals V_1 and V_2, between which we wish to measure the phase difference are connected. Again, the null detector is set to zero, but now by adjusting A only. The setting of A will then indicate the phase difference of the two signals. The accuracy of this method depends almost entirely on the accuracy with which the phase shift of A is known. This method is normally used for high-frequency signals.

By far the most popular method for measuring a phase difference is based on conversion of a time interval to a voltage. This type of phase meter first converts both input voltages V_1 and V_2 into rectangular waves by means of two comparators. These rectangular voltage signals are denoted V_1' and V_2' in Fig. 4.7. The time difference Δt between two positive edges of these signals is measured by a phase detector, which produces a pulse with a fixed known height V_p and the same width Δt. The average value V_{avg} of these pulses is determined by a low-pass filter. If T is the period of the input signals, this filtered voltage is equal to $V_{\text{avg}} = V_p \Delta t/T$. Therefore, a phase angle $\phi = 2\pi \Delta t/T$ amounts to:

$$\phi = 2\pi \frac{V_{\text{avg}}}{V_p}$$

The frequency range of this method for measuring phase differences is limited by the low-pass filter for low frequencies. The high-frequency limit depends on the speed of the logic circuits. The accuracy of the measurement depends on the accuracy of the voltage V_p, the difference in time delay of the two inputs sections of the phase detector and on the uncertainty with which the comparators detect a zero-crossing. This uncertainty is a result of the difference in offset of the two comparators. Due to this offset, the precise moment of a transition in the output of the comparator will depend on the slope of the input signal.

The above method of measurement, in fact, measures the time between two zero-crossings of the input signals. Therefore, the waveform of the input signal will have no effect on the measurement, provided that the zero-crossings remain in the same place.

Often we wish to measure the phase difference between two sinusoidal signals or, when the input signals are distorted, between the two fundamental harmonics. Distortion will generally give rise to measurement errors, as the presence of higher harmonics may shift the zero-crossing of a signal, or even produce extra zero-crossings. The harmonic distortion d_n of the nth harmonic, with an amplitude a_n, is defined as:

Figure 4.6. Phase compensation measurement method.

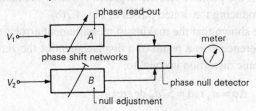

Figure 4.7. (a) Electronic phase meter using a time-interval-to-voltage converter. (b) Various wave shapes in this phase meter as a function in time.

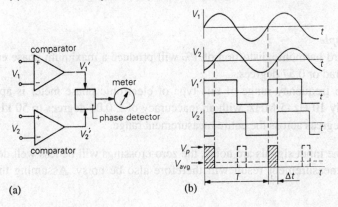

(a) (b)

$$d_n = \frac{a_n}{a_1} \quad (n = 2, 3 \ldots)$$

where a_1 is the amplitude of the fundamental harmonic. If $d_n \leq 1/n$ there will not be extra zero-crossings. This can be verified graphically. The permissible level of distortion is so high that we can safely assume that in a practical situation no extra zero-crossings will occur. If this were the case, however, excessively large measurement errors would result.

If the distortion is smaller than indicated, it will generally cause a shift of the zero-crossings with respect to those of the fundamental harmonic. Only when the harmonic is in phase or in antiphase with the fundamental, will this not occur. When the harmonics have a different phase, we must examine the even and odd harmonics separately. Even harmonics will cause a shift of the positive and negative zero-crossing in opposite directions, whereas odd harmonics will cause the same shift for all zero-crossings. It is therefore possible to compensate a shift which results from even harmonic distortion, but not a shift resulting from odd harmonic distortion. In the case of even harmonic distortion, the phase difference between two signals is measured once for the positive zero-crossing and once for the negative zero-crossings. Compensation is accomplished by taking the average of these two measurements. This can easily be achieved by triggering the phase detector in Fig. 4.7 on both the positive and the negative edges of the rectangular waves V_1' and V_2', producing the dotted pulses in Fig. 4.7(b).

It can be shown that the maximum measurement error $\Delta\phi$ of the measured phase difference ϕ as a result of a displacement of the zero-crossing due to odd harmonic distortion is equal to:

$$\Delta\phi = d_n \text{ rad} = 57d_n \text{ degrees}$$

This expression also applies to the maximum of the measurement error caused by even harmonic distortion, when the above compensation is not used.

Example
A third harmonic distortion of 1% will produce a maximum phase error of 10^{-2} rad or 0.57 degrees.

The frequency range of this type of electronic phase meter is approximately 10 Hz–5 MHz, with an inaccuracy of ± 0.05 degrees ≤ 50 kHz and ± 3 degrees across the entire measurement range.

If the input signals are noisy, the zero-crossings will be less well defined. The measurement result will therefore also be noisy. Assuming that, on average, the noise shifts the zero-crossings as much forward as backward, the influence of the noise of the measurement result can be reduced by taking the average value over a large number of measurements. The influence of the noise is also less when the slopes of the input signal zero-crossings are steeper. Ideal signals for measuring the phase difference are therefore rectangular wave signals.

4.3 Digital voltmeters

A digital voltmeter can, in fact, only measure a DC voltage. However, such a measurement can be performed extremely accurately (inaccuracy $\geq 10^{-5}$) and quickly (up to 100 measurements a second). This high measurement speed can be utilised for quantifying slowly varying AC signals. One can also place a scanner at the input of the digital voltmeter, which scans a large number of test points sequentially (time multiplexing, Fig. 2.10). Digital voltmeters are usually capable of communicating directly with a computer via for instance a GPIB output bus (which we will discuss in Section 4.5.7). Also, they are often programmable which allows the computer to give instructions such as when to start a measurement, which measurement range to use, etc. This makes it possible to realise a totally automatic measurement system. In addition to external programming, a digital voltmeter is also capable of automatically finding the correct measurement range (auto ranging) and even the polarity of the input signal (auto polarity). It is important that such a voltmeter uses a certain hysteresis for auto ranging, to avoid jumping back and forth between two adjacent ranges, when displaying a certain (nearly constant) voltage at the end of a range. If, for instance, the voltmeter is set to the 1 V range and the input voltage gradually rises, the meter will not switch to the 10 V range until the input passes 1.2 V (20% over range). Conversely, if the meter is in the 10 V range and the input voltage drops, the voltmeter will switch to the 1 V range when the input falls below 1 V.

Sometimes a digital voltmeter is equipped to measure other quantities as well as voltages. Such an instrument is referred to as a digital multimeter. These meters measure AC voltages with one of the methods for converting an AC signal into a DC voltage discussed in Section 3.3.5. The characteristics of the converter raise the inaccuracy to approximately 10^{-3}. Another function available in multimeters is resistance measurement. Resistance is measured by sending an accurately known current through the resistor and measuring the resulting voltage.

Digital meters are excellently suited for *accurate* and *quick* measurements and can easily collect *large numbers* of measurement data. For these applications an analogue display would be very tiring, time consuming and could easily give a large number of reading errors. If automatic measurements are required, the digital meter must be programmable and have a bus output.

A Digital VoltMeter (DVM) is designed around an ADC with a single measurement range (usually 10 V full-scale). Obviously, a digital voltmeter must have many input ranges, for instance from 100 mV to 1 kV, in steps of 10 (unlike analogue meters, the range switches of which have steps of 3). These input ranges are made by amplifying, low-level signals and attenuating high-level signals. If necessary, the polarity of the input signal can be changed to accommodate a single polarity ADC by means of an inverting amplifier.

The input amplifier, attenuator, polarity inverter and the ADC of the digital voltmeter are normally floating, i.e. free from ground, as is the case with an instrumentation amplifier (see Section 3.3.4). This is done to obtain a high rejection ratio of low-frequency common-mode (CM) voltages such as line hum, etc. The floating section of the meter is enclosed by a guard, which is connected to the CM voltage to minimise capacitive and inductive injection of interference. The digital control signals and the digital output signal of the meter are fed through the guard enclosure by means of small pulse transformers or opto-couplers.

Besides common-mode rejection, the term *normal-mode rejection* also exists. A variety of sources exhibit interfering AC signals superimposed on the DC voltage we wish to measure. This signal almost always consists of 50 Hz hum and higher harmonics. It can be treated as if it is connected in series with the ordinary input signal. The latter is called the normal-mode signal. A measure of the insensitivity of the DVM to this kind of disturbance is the Normal-Mode Rejection Ratio (NMRR). Normal-mode signals can cause large measurement errors. These errors can be reduced by positioning a low-pass filter in front of the DVM. Such an input filter is effective against noise, sharp voltage spokes, etc. in the input signal. However, for this filter to be effective against 50 Hz line hum, the bandwidth must be small (no more than 10 Hz); the digital voltmeter would become slow. A step input would take a relatively long period (approximately 0.1 s) before settling in the output of the low-pass filter. Only then can the DVM measure the step; so the overall response time would be longer still since it will include the conversion time of the DVM's ADC. Fortunately, there are also ADCs which are inherently insensitive to these power line related normal-mode disturbances. These ADCs are made to integrate the input signal for a certain length of time T and produce a digital output proportional to the average value of the input over the time T. In this manner, a disturbing periodic AC component whose fundamental frequency is equal to $1/T$ can be completely removed.

The NMRR is defined as the ratio of the peak value $\hat{v}$ of a sinusoidal normal-mode AC voltage $\hat{v} \sin \omega t$ and the maximum error ΔV_{max} of the displayed value, caused by this interference, so:

$$\text{NMRR} = \left| \frac{\hat{v}}{\Delta V_{max}} \right|$$

The NMRR is frequency dependent. For an integrating ADC with integration time T the frequency dependence can easily be calculated. If we connect only the disturbing normal-mode voltage signal to the input, we calculate for the average value over the interval $(t, t + T)$ the maximum output voltage:

$$\Delta V = \frac{1}{T} \int_{t}^{t+T} \hat{v} \sin \omega t \, dt = -\frac{\hat{v}}{\omega T} \{\cos \omega(t + T) - \cos \omega t\}$$

With a little trigonometry this can be rewritten as:

$$\Delta V = \frac{2\hat{v}}{\omega T} \sin \left(\tfrac{1}{2} \omega t \right) \sin \left(\omega t + \tfrac{1}{2} \omega T \right)$$

We can find the maximum value ΔV_{max} of the voltage by starting the AD conversion precisely when ΔV is at its maximum, so when $\sin (\omega t + \tfrac{1}{2} \omega T) = 1$. Thus:

$$\Delta V_{max} = \frac{2\hat{v}}{\omega T} \sin \tfrac{1}{2} \omega T = \frac{\hat{v}}{\pi f T} \sin \pi f T$$

and the NMRR is:

$$\text{NMRR} = \left| \frac{\hat{v}}{\Delta V_{max}} \right| = \left| \frac{\pi f T}{\sin \pi f T} \right|$$

In Fig. 4.8 the NMRR is plotted as a function of the frequency of a sinusoidally disturbing AC voltage for an ADC with an integration time of $T = 100$ ms. The function describing the NMRR has a pole for every $f = n/T$ (n is an integer). The lower limit of the NMRR is unity as f approaches 0 Hz. The dashed line tangent to the NMRR function is given by NMRR = $\pi f T$. Fig. 4.8 shows that the influence of disturbances is zero if the integration time *T* is chosen to be exactly equal to the repetition frequency of the disturbance or to a multiple (*n*) thereof. Since by far the most normal-mode interference is caused by the 50 Hz mains supply, *T* is usually chosen to be equal to *n* × 20 ms. Even if the line hum interference signal is distorted and contains harmonics, the harmonics are also completely eliminated if the integration time *T* of the DVM is chosen to be equal to (or equal to a multiple of) the period of the fundamental frequency of a periodic normal-mode interference signal.

The measurement rate of a digital voltmeter is lower than the maximum conversion rate of its internal ADC, since some time is also required for taking decisions on the polarity of the signal, the measurement range, zero-offset correction, etc. Therefore, the maximum number of measurements per second of the DVM is usually stated in the specifications. For an integrating ADC adjusting the measurement rate will only vary the delay between two measurements rather than the integration time, which is determined by the line frequency and held constant.

Whether a particular AD conversion method is suitable for a DVM depends largely on the following factors:
- accuracy,
- resolution,
- insensitivity to disturbance NMRR noise,
- conversion rate.

The resolution must be larger than required by the desired accuracy, as otherwise, this accuracy cannot be obtained. If, for example, the inaccuracy is equal to 10^{-4}, the resolution must at least be 10^4 to make full use of this inaccuracy. The display must therefore have at least four digits.

If the resolution is far better than required in view of the inherent accuracy, the digital voltmeter may be used for the accurate comparison of

Figure 4.8. The NMRR as a function of the frequency of an interfering normal-mode sinusoidal voltage for an ADC with an integration time of 100 ms.

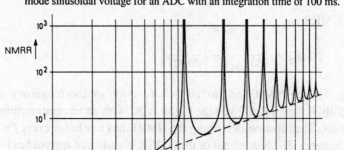

two quantities (substitution method). The NMRR affects the conversion rate; an integrating ADC is slower than a non-integrating ADC.

Fig. 4.9 shows a method for realising an integrating ADC which converts the input voltage into a frequency. First, the input voltage V_i is converted into a current I by a resistor with an accurately known capacitance. So $I = V_i/R$. This current is used to charge a capacitor with an accurately known capacitance C. The voltage V_c across the capacitor is then:

$$V_c = \frac{1}{C} \int_0^t I \, dt + V_c(0) = \frac{1}{RC} \int_0^1 V_i \, dt + V_c(0)$$

As soon as the voltage across the capacitor reaches a certain well-defined reference voltage V_R, the + comparator reverses the current, so the capacitor is now discharged but at the same rate. In this way, it is as if the slope of the capacitor voltage signal is folded to fit exactly between the two levels $+V_R$ and $-V_R$. This gives a sawtooth voltage with a peak-to-peak value of $2V_R$. The charge q_p which is supplied to the capacitor during the rising slope of the sawtooth is given by:

$$q_p = 2 \, CV_R = \int_{t_i}^{t_{i+1}} I \, dt = \frac{1}{R} \int_{t_i}^{t_{i+1}} V_i \, dt$$

Figure 4.9. An integrating ADC based on voltage-to-frequency conversion.

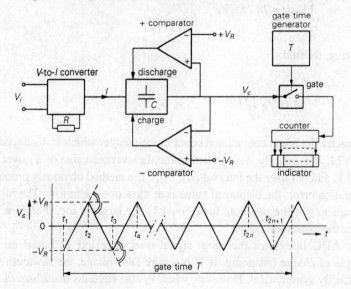

in which t_i and t_{i+1} are the points at which the positive-going slope starts and ends. During the negative slope of the sawtooth, a charge q_n is drained from the capacitor. This charge amounts to:

$$q_n = 2\,CV_R = \int_{t_{i-1}}^{t_i} I\,dt = \frac{1}{R} \int_{t_{i-1}}^{t_i} V_i\,dt$$

The points t_{i-1} and t_i coincide with the beginning and end of the negative slope of the sawtooth wave. Obviously:

$$q_p = q_n \quad (charge\ balance)$$

and therefore also:

$$\int_{t_{i-1}}^{t_i} V_i\,dt = \int_{t_i}^{t_{i+1}} V_i\,dt = 2\,RCV_R$$

When the sawtooth reaches a minimum (or a maximum), the gate is opened and a counter starts counting the number of complete periods of the sawtooth. The gate is opened for a fixed known gate time T. If the total number of complete periods of the sawtooth fitting in the interval T is denoted by n, we can write:

$$\int_{t_1}^{t_{2n+1}} V_i\,dt = 2n\,2RCV_R$$

Rewriting, we find:

$$n = \frac{1}{4RCV_R} \int_{t_1}^{t_{2n+1}} V_i\,dt$$

n is exactly the total accumulated count in the counter which is displayed by the DVM. We actually wish to determine the average value of V_i over the period T, and not over the interval $t_{2n+1} - t_1$. The method obviously produces a truncation error; the displayed value is at most one count low. The higher the frequency of the sawtooth for a given V_i, the larger n and, therefore, the smaller the truncation error.

This ADC integrates the input signal over a period T, based on the principle of charge balancing. If V_i is a pure DC voltage, the sawtooth will be perfectly symmetrical. However, when V_i also contains disturbances, the waveform will no longer be symmetrical and the momentary frequency will vary. The NMRR depends on the length of the integration time T. The accuracy of the measurement depends on the accuracy of R, C, the reference voltages $+V_R$ and $-V_R$ and the accuracy of T. As expected, the major disadvantage of this integrating voltage-to-frequency conversion type of ADC is that its correct operation relies on the accuracy of too many components.

We can determine the values that one needs for the components of this type of ADC as follows: Assuming the required input range is 0 to +10 V with a maximum inaccuracy of $\pm 5 \times 10^{-4}$, a resolution of 10^4 is then definitely necessary. The integration time T must be a multiple of 20 ms to produce a high NMRR for the 50 Hz hum. A final requirement is a maximum measuring rate of 10 measurements per second. The longest allowable integration time is then equal to 80 ms. Over this time, $V_i = 10$ V must be displayed as $n = 10^4$ in the counter to yield a resolution of 10^4. For the DC voltage V_i, $n = TV_i/4RCV_R$, from which it follows that $RCV_R = 2 \times 10^{-5}$ V s. If we choose $V_R = 5$ V the peak-to-peak value of the sawtooth wave will become 10 V which is easy to realise. Practical values for R and C are $R = 4$ kΩ and $C = 1$ nF. If T, R, C and V_R are manufactured with a relative inaccuracy less than 10^{-4}, the specification for accuracy is met. With $V_i = 10$ V, the frequency of the sawtooth becomes $n/T = 125$ kHz. With lower input levels this frequency is proportionally lower.

The number of accurate components needed can be reduced by integrating a known reference voltage just before or after integrating the measured input signal. The ratio of both measurement results will no longer contain the values of the circuit components (as long as they do not change during the measurement). The accuracy of this *voltage ratio measurement* is predominantly determined by the accuracy of the reference voltage and the linearity of the ADC. This *ratio method* is utilised in the so-called dual-slope integrating ADC. Fig. 4.10 gives an example of a dual-slope integrating ADC.

During the first half of the measurement, the input voltage V_i is integrated over a constant integration time T, for instance by a voltage-to-current converter ($I = V_i/R$) and a capacitor, or by a Miller integrator (the capacitor C is then connected as negative feedback across an operational amplifier). The voltage V_c across the capacitor, at the end of the integration period T, will amount to:

$$V_c(t_1 + T) = \frac{1}{C} \int_{t_1}^{t_1+T} I \, dt = \frac{1}{RC} \int_{t_1}^{t_1+T} V_i \, dt$$

Evidently, this voltage is proportional to V_i. Also, at the beginning of the integration period, i.e. at $t = t_1$, the counter is set to zero. The integration continues until the counter overflows (count equals 10^n if the required resolution is 10^n). Therefore, if f_0 is the frequency of the reference oscillator, the integration will last for a time $T = 10^n/f_0$. At $t_1 + T$ the counter overflow bit switches the input of the ADC to the reference voltage V_R, which has a polarity opposite to V_i. This causes the voltage across the capacitor to fall now at a constant rate of $dV_c/dt = -V_R/RC$. A comparator detects when V_c becomes zero and stops the measurement. The reading of the counter at this time t_2 will equal:

$$f_0 \{t_2 - (t_1 + T)\} = f_0 \frac{V_c(t_1 + T)}{V_R/RC} = \frac{f_0}{V_R} \int_{t_1}^{t_1+T} V_i \, dt$$

For the DC component (time average) of V_i the displayed count is:

$$\frac{f_0}{V_R} V_i T = \frac{V_i}{V_R} 10^n$$

Therefore, the measurement gives a number proportional to the *ratio* of V_i and V_R. The full-scale value of this dual-slope integrating ADC is determined by the magnitude of V_R. The only requirement for R, C and the frequency f_0 is that they have a sufficient short-term stability not to exhibit any drift during the time required to complete the two cycles of the conversion. The accuracy of the components no longer plays a role. This method

Figure 4.10. The principle of a dual-slope integrating ADC.

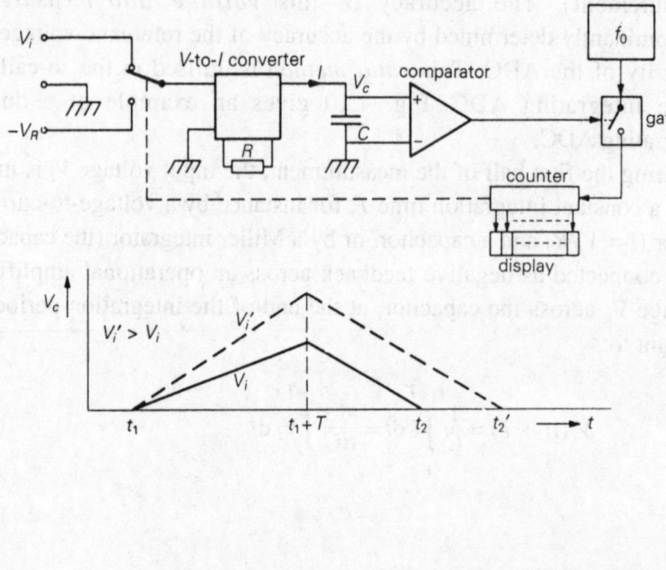

also makes use of the charge balance principle, since the supplied charge during the first integration is equal to the drained charge during the second integration. The maximum number of measurements per second is small, as two integrations must be performed and time is also lost in switching, resetting, etc.

This disadvantage is eliminated by using the *two-system principle*. A compensating ADC is fast and accurate, but sensitive to disturbances. An integrating ADC on the other hand is insensitive to disturbances, but slow. The two-system principle combines the favourable characteristics of the compensating and the integrating ADCs as follows: A DAC is used to compensate the input voltage coarsely. The remaining difference voltage contains all the disturbances. It is measured by means of an integrating ADC. If this ADC has a relative inaccuracy ε, and the compensation is complete but a fraction δ, and the DAC has a relative inaccuracy γ, the total inaccuracy will amount to $\gamma + \varepsilon\delta$.

Assume that the DAC has an inaccuracy of $\gamma = 10^{-5}$, the inaccuracy of the integrating ADC is $\varepsilon = 10^{-3}$ and the compensation is accomplished up to 1%, so $\delta = 10^{-2}$. The total error will then be equal to 2×10^{-5}. If the resolution is 10^5 the two most significant decimals of the display are determined by compensation and the three smaller decimals by the integrating ADC. Thus, the resolution of the latter may be small. If, for instance, an ADC based on voltage-to-frequency conversion were used the highest frequency of the sawtooth voltage necessary to obtain an integration time of 20 ms would be 50 kHz. However, if this ADC were to produce all five digits of the display, its highest frequency would have to be 5 MHz to satisfy the same integration time. If the highest frequency were to remain at 50 kHz, the integration would take 2 s. Frequencies higher than 100 kHz reduce the accuracy of the voltage-to-frequency conversion drastically, due to the non-ideal high-frequency characteristics of its components. Therefore, a two-system ADC operating with partial compensation of the input voltage is quick, accurate and insensitive to disturbances.

4.4 Oscilloscopes

An electronic instrument which is very often used to measure electrical wave forms is the oscilloscope (L. *oscillare* = to swing; Gr. *skopeein* = to view). The function of an oscilloscope is to make an electrical signal $y(t)$ visible as a function of time by faithfully reproducing it on the screen of a CRT. This is achieved by displaying the signal $y(t)$ concurrently against a signal $x(t) = ct$ generated internally in the oscilloscope. The displayed result is $y = y(x)$. A new time base for the signal $y(t)$ is created by the instantaneous amplitude of the signal $x(t)$. Since $y(x) = y(ct)$ the adjustable constant c makes it possible to expand or compress the time base. It will then appear as if y varies increasingly slower or faster. An oscilloscope needs a time base signal generator to create $x(t)$ and an analogue two-dimensional display which allows y to be displayed as a function of x. The limited physical dimensions of the display imply that only a short interval of $y(t)$ can be made visible. However, if $y(t)$ is a periodic signal with period T, an apparently continuous image of $y(t)$ can be obtained by repeatedly displaying $y(t + nT)$, provided that the time base signal starts for every n at the same phase of $y(t + nT)$. For this, we need a circuit which can produce a trigger signal to start $x(t)$ when a given phase of $y(t + nT)$ is detected.

In order to allow a large flexibility for analysing a wide variety of waveforms, oscilloscopes are sometimes implemented in the form of a so-called main frame into which various plug-in units can be inserted. The main frame contains the actual oscilloscope with the CRT as an analogue (x,y)-display, a time base signal generator, a trigger circuit and a power supply unit. The plug-in units offer the possibility of measuring various parameters of the input signal, such as amplitude, frequency, time interval, frequency spectrum, etc.

The internal organisation of an oscilloscope is illustrated in Fig. 4.11. The CRT is the heart of the oscilloscope. This device has already been discussed in Section 3.4.2. The signal to be visualised is supplied via the y-channel to the vertical deflection plates of the CRT. This y-channel provides amplification (small signals) and attenuation (large signals) and sometimes adjustable filtering (noisy signals). In high-frequency oscilloscopes (above

Figure 4.11. Functional organisation of an oscilloscope.

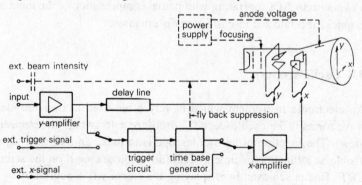

25 MHz) a delay line (a coil of co-axial cable or a number of LC sections) is inserted in the y-channel. This delay line is necessary for making the leading edge of signal pulses visible. This makes it appear that the time base starts to write the y-signal on the CRT screen *before* the oscilloscope triggers. In this way the delay of the trigger circuit is cancelled. Also, for very rapidly rising signals, the portion of the signal directly ahead of the trigger point is made visible.

The x-signal can be supplied externally, via the deflection amplifier of the x-channel. In this manner, it is possible to create Lissajous figures for phase measurement. The time base signal is generated internally by integrating a constant current resulting in a sawtooth-shaped deflection voltage. The starting signal for the time base is given by a trigger generator, which derives the trigger signal either from the y-signal or from an external input.

A beam blanking signal is derived from the time base to suppress the electron beam of the CRT during the retrace of the beam. Often it is also possible to modulate the intensity of the electron beam by applying an external signal (z axis modulation). We will now examine the most important functions of an oscilloscope.

Time base
In Fig. 4.12(a) the horizontal deflection voltage V_x is plotted, as produced by the time base generator. Immediately after receiving a trigger pulse, the blanking of the electron beam is ended and the sawtooth waveform starts to rise. The electron beam now starts to sweep across the CRT screen displaying the measured signal. Once the beam reaches the end of the screen (V_x is at its maximum), it is cut off again and the time base voltage V_x is reset to its initial value. This suppresses the fly-back of the beam. The oscilloscope is

Figure 4.12. (a) Time base signals $V_x(t)$ for the horizontal deflection of the electron beam in the CRT of an oscilloscope. (b) Single time base signal in the single-sweep deflection mode. (c) Time base signal in the magnified time base mode. (d) Time base signal in the delayed time base mode.

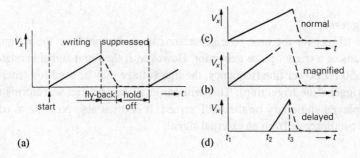

now ready to display the next cycle of the waveform; it is waiting for the following trigger pulse. Between beam sweeps, at least one period of the measured signal is passed over. For high-frequency signals a number of periods of the input signal are usually skipped before the oscilloscope is ready to write the next trace. The delay is necessary predominantly for resetting the time base generator. By adjusting the slope of the sawtooth wave form we can choose the length of the input signal shown on the screen. The horizontal scan time is varied, resulting in the visualisation of a longer or shorter interval of $y(t)$. The time base scan time may vary from 5 s/cm to 0.5 ns/cm for high-frequency oscilloscopes.

Most oscilloscopes offer the facility of increasing the gain of the x-amplifier by a factor of 5 or 10. This results in a horizontal magnification of the displayed signal around the centre of the screen. Any section of the input signal can be so magnified by adding a DC voltage to the x-deflection voltage. This will vary the horizontal position of the signal along the screen. Fig. 4.12(c) shows, for example, the deflection voltage V_x which will expand the first part of the input signal, directly after the trigger point. However, this method of expanding the time base also magnifies the inaccuracies in the time base signal and the exact time between the trigger point and the shifted signal is lost.

A method which does not have this disadvantage is based on the creation of a *delayed sweep*. If a small time interval of the input signal, some time after the trigger point t_1 in Fig. 4.12(d), calls for detailed examination (for instance, there is a sharp transition between t_2 and t_3) then a delayed time base can prove useful. This time base is armed by the normal trigger circuit but is not started until after a delay, at time t_2. The delay $t_2 - t_1$ can be accurately set by means of a timer circuit. If the (delayed) sweep time is chosen so that the end of the sawtooth coincides with t_3 only the interval (t_2, t_3) is displayed, expanded across the full width of the screen. This way, a section of the signal at an accurate point in time after the trigger pulse can be examined in detail on an expanded time scale. This is especially useful for analysing impulse shaped signals.

Triggering
The trigger for the time base generator can be derived from the y-signal by means of a trigger pulse generator. However, if the input signal is related to the 50 Hz power line frequency, the line voltage may be used for generating a trigger (line triggering). This will cause all line related waveforms to be displayed stationary on the CRT screen. It is often also possible to trigger the oscilloscope from an external signal.

The trigger pulse generator compares the input signal voltage V_y to an adjustable DC voltage as illustrated in Fig. 4.13. The input signal will intersect this DC voltage during the rising edges of the signal and also during the falling edges. The trigger pulse generator can be set to produce only a trigger pulse on either a positive or a negative slope of the input signal. Adjustment of the trigger *level* and the trigger *slope* enables triggering at any point of the waveform in Fig. 4.13. For instance, if a trigger is required at t_1, the trigger level is set positive and the trigger slope is also set positive. To trigger at t_2, the trigger level and slope both have to be set negative.

Multichannel sscilloscope

If we wish to measure several signals simultaneously in order to examine their coherence in time, it is possible to use an oscilloscope with a CRT equipped with more than one electron gun and multiple vertical deflection plates. All electron beams are deflected by the same two horizontal deflection plates. Obviously, this is an expensive solution. It is simpler and less costly to apply the (*time*) *multiplex method* (see Section 2.2). However, the bandwidth of the CRT and the deflection amplifiers must be sufficiently large to allow this. Time multiplexing of a CRT can be implemented as follows: The trigger signal is derived from the input signal which has the steepest slope. After the time base triggers, the first signal is displayed. The y-amplifier now switches over to the second signal. This signal is displayed starting at the trigger derived from the first signal. The offset between the two traces is adjustable by applying a DC voltage to the vertical deflection plates. This mode of operation is referred to as the *alternating mode*. Unfortunately, it has two disadvantages. For low time base settings, the two traces tend to flicker. Furthermore, the relation in time between the two signals is not entirely preserved, as both traces are triggered by the first signal and the second trace is displayed some time after the first trace. These disadvantages can be avoided by switching back and forth very quickly (at

Figure 4.13. Adjustment of the trigger amplitude level and the trigger slope to enable triggering at any point of the input wave form $V_y(t)$.

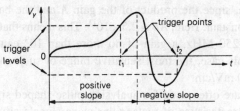

approximately 250 kHz) between the two input signals in the so-called *chopped mode*. Both traces are written simultaneously here. A piece of one trace is written during a gap in the other trace. If the time base is set low with respect to the chopping frequency, the pieces appear as if they are joined together; the traces look continuous.

Sampling oscilloscope

To display very high-frequency signals with an oscilloscope, the bandwidth of the y-amplifier, the delay line and the CRT must be extremely large. Fig. 3.74 shows that this easily leads to expensive solutions. However, even when the bandwidth of the oscilloscope is far too low to display these signals directly, it is still possible to use the oscilloscope provided that the input signals are periodic. This can be accomplished with coherent sampling (see Fig. 2.9). At every trigger pulse (i.e. every time the input signal exceeds a certain level) a single sample of the input signal is taken. The subsequent samples are shifted slightly in time with respect to the primary trigger. The effect of this is that each successive sample is taken a brief time interval δ later than the previous one. When the samples are displayed against a time base signal at a much lower frequency, the wave shape of the original high-frequency signal is faithfully restored. A 20 kHz sampling oscilloscope, for instance, is easily capable of displaying periodic input signals of up to 15 GHz!

Oscilloscope amplifiers

The signal which is to be made visible by the oscilloscope is not suitable for connecting directly to the deflection plates of the CRT, because the deflection sensitivity is too low and the capacitance of the deflection plates would load the measurement source. Therefore, an amplifier is placed between the y-input and the deflection plates. The gain of this amplifier is step-wise adjustable (*sensitivity*).

If the deflection sensitivity of the vertical deflection of a given CRT is 10 V/cm, the gain of the y-amplifier must amount to 5×10^5 in order to display small input signals with a sensitivity of 20 μV/cm. The available bandwidth is then not very large (500 kHz). For a large bandwidth, the gain cannot be chosen too large, since the product of the gain A and the bandwidth B is approximately constant. Here, $AB \approx 2.5 \times 10^{11}$. This means that if we require a bandwidth of 250 MHz, a gain of no more than 10^3 can be obtained. Therefore, for this case, the most sensitive range of such an oscilloscope would be only 10 mV/cm.

Oscilloscopes are often used for analysing pulse-shaped signals. In Fig. 4.14 various characteristics of rectangular wave signals are defined. The shape of the signal is influenced by the dynamic behaviour (the frequency dependence) of the system through which it flows.

Oscilloscope amplifiers are usually designed to amplify a pulse with as little distortion as possible. The overshoot and the ringing of the output signal are less than 2% for a step input. This implies, though, that the frequency characteristic is not maximally flat.

The trailing end of the flat top of the pulse contains information on the low-frequency behaviour of the transferring system, as indicated in Fig. 4.14(b). If the low-frequency transfer characteristics are poor, because the low cut-off frequency f_l of the bandwidth of the system is chosen too high, the top will not be flat, but slowly droop away. This droop is a measure for an inadequate low-frequency band end.

The rising and falling edges of the pulse contain information on the high-frequency behaviour of the system; it is characterised by finite rise and fall times. The rise time and the fall time are defined as the transit time required for the output to change between 10% and 90% of the total amplitude of a step variation. There is a relation between the high-frequency −3 dB point f_h and the magnitudes of the rise time τ_r and the fall time τ_f. In a linear system τ_r and τ_f are equal: $\tau = \tau_r = \tau_f$. The relation between f_h and τ is equal to:

$$f_h = \frac{c}{\tau}$$

A system whose dynamic behaviour is of the first order (e.g. an RC network) has a proportionality factor $c = 0.35$. (This can be proven with the aid of the discussion in Section 2.3.3.2). Almost all oscilloscopes which have a good step response and little overshoot or ringing have a proportionality factor:

$$0.32 < c < 0.4$$

Figure 4.14. (a) The input signal (dashed line) and output signal of an oscilloscope amplifier. (b) Characteristic quantities of a pulse-shaped output signal.

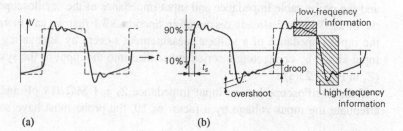

(a)   (b)

If an oscilloscope with an internal rise or fall time τ_0 displays a pulse with a rise and fall time τ_i, and $\tau_i \geq \tau_0$, the rise time of the displayed trace τ_t can be well approximated by the expression:

$$\tau_t = \sqrt{\tau_i^2 + \tau_0^2}$$

The rise time of the input signal is then:

$$\tau_i = \sqrt{\tau_t^2 - \tau_0^2}$$

This expression only holds true for a particular type of frequency response (a Gaussian frequency response), but it can also be used for other kinds of frequency response to a good approximation. If the above expression is not used to correct the observed τ_t, we find that for $\tau_i = 5\tau_0$ an error of 2% results and for $\tau_i = 7\tau_0$, the error is 1%. However, if τ_i is smaller than τ_0, large errors will remain, even after the observed time τ_t has been corrected with the above expression. So here correction is not possible. This is made clear by the following example. Assuming that the observed rise time τ_t has a relative error of $\Delta\tau_t/\tau_t = \pm 5\%$, then when $\tau_0 = 3\tau_i$, the resulting relative error in the calculated value of τ_i is $\Delta\tau_i/\tau_i = 50\%$! This can be found with: $\Delta\tau_i/\tau_i = \{1 - (\tau_0/\tau_t)^2\}\Delta\tau_t/\tau_t$.

Measurement probes

The signal source is usually not conveniently close enough to the (large) oscilloscope to be connected directly to the input; a probe is needed. There are voltage probes and current probes. A current probe consists of a ring-shaped ferrite core, through which the signal-carrying conductor is led. This conductor acts as the primary of a transformer. The secondary consists of several windings around the ferrite core. If we also wish to measure DC currents the principle of Fig. 3.22 can be used.

As well as passive probes, active probes are used which have a built-in amplifier with a unity gain, especially for making high-frequency measurements (to prevent loading). The amplifier operates as a buffer amplifier between the source impedance of the measurement object on the one side and the probe cable impedance and input impedance of the oscilloscope on the other. We have already discussed in Section 3.3.1 that we can increase the input impedance of a voltage measurement system by attenuating the input signal by connecting a series impedance into the input of the system (Z_s in Fig. 3.23(a)).

If the oscilloscope has an input impedance $Z_i = 1$ MΩ//18 pF and we attenuate the input voltage by a factor of 10, the probe must have series impedance of $Z_s = 9$ MΩ//2 pF. Thus, the input impedance of the total system becomes $Z_i = 10$ MΩ//1.8 pF and is considerably increased from the original 1 mΩ//18 pF at the expense of a reduced (× 10) sensitivity.

The impedance of the probe must be adjusted to obtain a frequency independent attenuation. This can be done by applying a square wave signal to the probe and adjusting a small variable capacitor located in parallel to the series resistance of the probe. In this manner, the effect of the input capacitance of the oscilloscope on the attenuation can be eliminated. When the edges of the square wave on the CRT are still rounded, the frequency correction is not large enough. When the edges exhibit peaks the correction is too large. The frequency correction capacitor must be adjusted to produce the most square signal on the CRT screen.

Accuracy

The accuracy of oscilloscope amplifiers and attenuators is matched to that of the CRT. As a consequence of the finite dimensions of the screen, the diameter of the light spot (thickness of the trace) and the non-linearity of the deflection, an accuracy of 1% is more than adequate. The total inaccuracy of the oscilloscope is often no better than ± 3%. The inaccuracy of the time base of the oscilloscope is also approximately ± 3%. An interval measurement may be performed a little more accurately using the delayed time base (± 1%).

4.5 Data acquisition systems

Many measurement systems currently make use of digital computers. The reason for this is that digital computers (as opposed to analogue computers) have become less expensive, faster, smaller, and markedly more reliable in recent years. In addition, a digital computer can add tremendous flexibility to a measurement system, since its software can be readily modified by the user. All these factors contribute to the reason why micro and mini computers are used in many measurement systems today. Examples range from fully automatic testing of end-products in industry, to automatic controlling of industrial processes, to sophisticated automation in laboratories. In all these applications not only the computer's ability to control the various functions of the measurement instrumentation, but also its ability to perform a wide spectrum of operations on the measurement information plays a major role. Consider, for instance, operations to reduce noise and distortion in measured data, operations to correct for non-linearities and disturbances, and operations to process the measurement data so that it is easily inter-preted by humans.

The overall system required for the capturing, processing, displaying and distribution of measurement information is generally called a Data Acquisition (and Distribution) System (DAS). In the subsequent sections of this chapter we shall examine various aspects of a DAS.

4.5.1 Introduction

We have already seen (Section 3.3.6) that virtually all measurable parameters and variables of macroscopic physical processes in the world around us are analogue in nature. We therefore define an analogue measure-ment signal as a signal whose value (amplitude) is defined everywhere on a specific (time) interval, and whose value may assume all values between specific upper and lower limits. An analogue signal is therefore a signal continuous in the time domain as well as in the amplitude domain.

Digital computers cannot manipulate these signals directly; they can only act on and produce digital signals. To go from analogue to digital signals (and vice versa) we need a type of processing called 'signal conversion'. Part of the signal conversion is the mapping of a time-continuous signal into a time-discrete signal. A time-discrete signal is only defined at specific points in time. This discretisation is necessary since computers cannot process data continuously in time but only at discrete points in time. A time-discrete signal that follows from an analogue signal by (only) making the time domain discrete is known as a 'sampled data' signal. Such a signal may easily be produced with an electronic sampling circuit.

In addition to processing at discrete points in time, a computer has only a limited word width. It can therefore only ingest the signal amplitude with a finite resolution. The computer is discrete not only in the time domain, but is also quantised in the amplitude domain. Therefore, the amplitude of the input signal must be made to change only in finite steps. This is referred to as 'amplitude quantisation'. This quantisation is realised with an ADC (see also Section 3.3.6). Before it can be entered into a computer, a measurement signal must therefore be sampled (in time) as well as quantised (in amplitude). This type of signal, which is time-discrete and has a quantised amplitude, is called a 'digital signal'.

In the following sections we shall see that, before sampling, an analogue signal must be filtered to remove the higher-frequency components (noise and distortion). These would otherwise lead to errors. This prefiltering is an essential part of the Data Acquisition and Distribution System (DAS). This is not the case for the processing needed to give the signal an amplitude level and a waveform optimal for the further processing by a DAS. The latter type of processing is known as 'signal conditioning'. Fig. 4.15 shows the various sections of a DAS along with the different types of signal in these sections. The 'D' for 'Distribution' is usually left out in a Data Acquisition and Distribution System (DAS), even if the system is used to output analogue drive and control signals. In order to output signals, the digital output signal from the computer in a DAS is first transformed into an output voltage or current. This is done with a DAC (see Section 3.3.6) which translates the digital word D_j into a proportional voltage or current. The amplitude of the output voltage or current is then only determined at discrete points t_i in time. To arrive at an analogue output again, interpolation is used. The interpolation of the output amplitude between the points t_i is known as 'reconstruction'. This is usually performed by a low-pass filter: the so-called 'reconstruction filter'. In practice the processing capacity of the digital computer is often so large that more than one input signal can be processed, using time multiplexing (see Section 2.2). In this case a 'multiplexer', or 'scanner', is necessary at the input side, while a 'demultiplexer', or 'distributor', is placed at the output side of the computer. The DAS in Fig. 4.15 is a single-channel DAS and therefore does not require multiplexing and demultiplexing components.

Figure 4.15. DAS comprising only one single measurement channel.

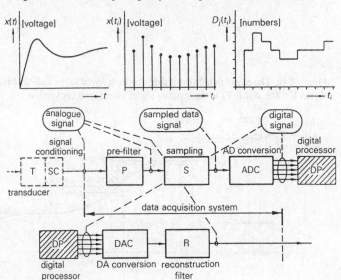

4.5.2 Digitisation

We have seen in the previous section that a DAS must digitise the analogue signal at the input side of the processor; it has to convert the signal such that it is discrete in time as well as quantised in amplitude. To do this, the analogue signal must be converted at sampling points t_i into a stream of binary words D_j. These words are usually fed in parallel to the processor, with one complete word per sampling point. The conversion of an analogue signal into a sequence of digital words introduces errors and distortion. Fig. 4.16 shows these errors in the time and amplitude domain for the analogue input signal $V_A(t)$. Due to the uncertainty Δt in the sampling points t_i we create an amplitude error ΔV which depends on the 'steepness' of the signal $V_A(t)$. These errors become larger as the high-frequency content of $V_A(t)$ increases. If there were no error at all, the binary signal D, restored to the domain of the input signal, would appear as V_A' in Fig. 4.16. In reality V_A' varies within the band indicated by dashed lines in Fig. 4.16.

The error ΔV_t then becomes:

$$\Delta V_t = |V_A' - V_A|$$

This (total) error ΔV_t is larger than ΔV. This is due to quantisation errors, inaccuracies in the AD conversion, and sampling errors on top of the sampling timing errors. In the following sections, we shall discuss the errors which arise from quantisation and sampling.

Figure 4.16. The errors in the signal conversion, referred to the analogue input signal V_A. The signal V_A' is the restored signal after signal conversion.

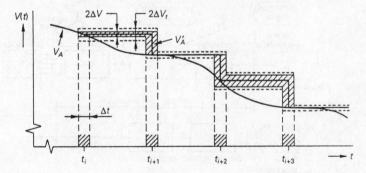

4.5.3 Quantisation theory

Even if we were to assume that the sampling and the AD conversion were ideal (i.e. without error), we would still have to deal with the quantisation errors (see Section 3.3.6). In Fig. 3.63(b) this error is illustrated for a linearly increasing input signal V_A. The quantisation error Q is thus:

$$Q = \Delta V_t = V_A' - V_A$$

where V_A' is the restored value of the binary output of the ADC. Since this binary word $D = (a_n a_{n-1} \ldots a_0)$ is comprised of n+1 bits, the smallest increment V_0 is:

$$V_0 = \frac{V_R}{2^{n+1} - 1}$$

The ADC of Fig. 3.63 produces a so-called 'round-off quantisation error'. With this type of quantisation, the maximum possible error is:

$$-\frac{1}{2} V_0 \leq Q \leq \frac{1}{2} V_0$$

Round-off errors therefore give a maximum quantisation error of plus or minus one half the Least Significant Bit (LSB). The quantisation error which is made by a truncation lies within the interval:

$$0 \leq Q \leq V_0$$

We can lump the quantisation effects in the signal path of a DAS together, and represent them by a so-called 'quantiser'. In the above expressions, we have given a non-linear model for this quantiser for both round-off and truncation. In Fig. 4.17 this is illustrated once again for a round-off error. The problem with this approach is that this non-linearity is difficult to describe analytically. Therefore we present a stochastic model in Fig. 4.17(d). Here we view the quantiser as a source of additive quantisation noise; the output signal is equal to the sum of the input signal V_A and the quantisation noise V_N. This quantisation noise has an amplitude probability density function $f(Q)$ as is depicted in Fig. 4.17(e). Q assumes all values between $+\frac{1}{2} V_0$ and $-\frac{1}{2} V_0$ with equal probability; the probability density function is uniform. The mean value $\overline{Q} = 0$ and the variance σ_Q^2 is given by:

$$\sigma_Q^2 = \int_{-\infty}^{\infty} (Q - \overline{Q})^2 f(Q) \, dQ$$

which equals:

$$\sigma_Q^2 = \frac{V_0^2}{12}$$

The RMS value V_N (standard deviation) of the added noise signal is therefore:

$$V_N = \sigma_Q = \frac{V_0}{2\sqrt{3}} = \frac{V_R}{2\sqrt{3}(2^{n+1} - 1)}$$

The Signal-to-Noise Ratio (SNR) resulting from this quantisation error, for a sinusoidal input signal with a peak value αV_R ($0 \le \alpha \le 1$), can be calculated as:

$$SNR = 6\alpha^2(2^{n+1} - 1)^2$$

This expression is plotted in Fig. 4.18, using a logarithmic scale (dB) for the SNR. It is apparent from this plot that it is important to drive the ADC with the largest possible input signal (just so no clipping occurs). Also, we see that the SNR increases by 6 dB for every added bit of ADC resolution.

The quantisation noise of the ADC in a DAS is not the only source of quantisation error. The digital processor processes the signal with only a finite word width. Therefore, this also introduces quantisation errors (which can become quite large when a 'fixed-point' processor is used). This is most easily shown with the example of a simple digital multiplication. A digital

Figure 4.17. The quantiser as the source of quantisation noise in the signal path of a DAS. (a) The quantiser as a non-linear element with characteristic (b) and quantisation error Q (shown in (c)). (d) Stochastic model of a quantiser. The added noise V_N has the amplitude probability density function shown in (e).

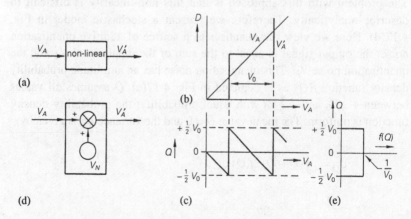

input signal D_1 input to the processor comes from an ADC and has a quantisation error Q_1 (standard deviation σ_1). In the processor, this signal is multiplied by the number D_2 which is obtained from a calculation, and therefore has a limited word width as well. This number contains the quantisation error Q_2 (standard deviation σ_2). The result of the multiplication is a number D_3:

$$D_3 = (D_1 + Q_1)(D_2 + Q_2)$$

If we assume that the two quantisation noise sources are not correlated with each other, nor with the signals D_1 and D_2, we find by using Gauss' propagation rule (see Section 2.3.2):

$$\sigma_3^2 = (D_2 + Q_2)^2\sigma_1^2 + (D_1 + Q_1)^2\sigma_2^2$$

Ignoring the smaller terms, we find:

$$\sigma_3^2 = D_2^2\sigma_1^2 + D_1^2\sigma_2^2$$

The result of this multiplication will be output with a limited word width. This truncation error gives quantisation noise with a standard deviation σ_4 at the output of the multiplier. Since this noise source may be assumed to be independent of both previously mentioned sources, we can express the total variance σ_0^2 as:

$$\sigma_0^2 = \sigma_3^2 + \sigma_4^2 = D_2^2\sigma_1^2 + D_1^2\sigma_2^2 + \sigma_4^2$$

Figure 4.18. The effect of quantisation noise on the SNR for a sinusoidal signal with a maximum value equal to α times the full-scale value V_R of the ADC.

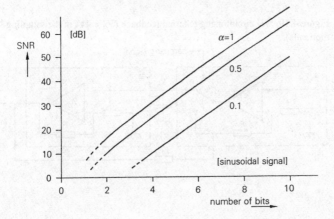

The word widths for each of the above variables need not be equal. The associated variances are equal to $(1/12)(LSB)^2$, since the respective probability density functions are uniform. It is therefore important to keep the number of multiplications as low as possible during the processing, and to choose to be the multipliers as close as possible to a realisable binary number.

In conclusion, we can say that quantisation in a DAS produces noise which:

– is added to the output of the quantising components;
– has a uniform amplitude probability density function;
– has a standard deviation (or RMS-value) equal to LSB/$\sqrt{12}$;
– is not correlated with the signal.

4.5.4 Sampling theory

Before we discuss the underlying theory of sampling, let us first examine a practical example of an electronic sampling circuit. The circuit will periodically freeze the input signal so that the ADC sees a constant input voltage during its conversion cycle. Such a circuit is known as a '*Sample and Hold*' (S/H) circuit. Fig. 4.19 shows a possible realisation of such an S/H circuit. When the sampling switch S is in position '0', the output voltage V_A' follows the input voltage V_A; the circuit is in the so-called 'tracking' mode. Since then, amplifier A_1 is an open-loop amplifier, and amplifier A_2 uses internal feedback such that $A_2 = 2$, the external feedback path reduces the gain to $V_A' = 4V_A$. In position '1' of the switch S, A_1 also contains an internal feedback path such that $A_1 = 2$. Capacitor C is charged to the voltage $2V_A$ at the instant when S is moved to position '1'. In this state the circuit is in the 'hold' mode; the output voltage $V_A' = 4V_A$, where V_A is the value of the input

Figure 4.19. S/H circuit using external feedback ($V_A' = 4V_A$ at the sampling moments).

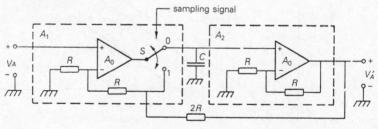

voltage at the time at which the switch was thrown to the lower position (sampling instant).

In Fig. 4.20 a sinusoidal input signal V_A and a sampling signal (the position of switch S in Fig. 4.19) are plotted, along with the resulting output signal V_A'. While $S = 0$ the output follows the input signal. When $S = 1$ the circuit holds the last value of V_A, from the instant at which S became 1. The input signal V_A is therefore sampled at the points in time where S switches from 0 to 1. For this reason, this specific circuit would be described more accurately by the term 'track and hold' circuit.

Unfortunately, an S/H circuit introduces time and amplitude errors. This is because the actual sampling point is later in time than the point at which the command is sent from the processor. In addition, the magnitude of the locked-in voltage at which the output is frozen during 'hold' differs some-what from the value of the input voltage at the actual sampling point (the rising edge of S in Fig. 4.20). These errors are illustrated in Fig. 4.21 in the form of a timing diagram.

After the 'track' command (S goes to '0'), there is a small time delay (the acquisition time) before the output signal $V_A'(t)$ again follows the input signal to within the transfer tolerance specified for the particular SH circuit. After this time delay, the circuit is ready to sample the input signal again. The acquisition time is therefore the minimal (idle) time that must be placed between a 'track' and a 'hold' command in order to avoid making excessive

Figure 4.20. Input signal V_A, sampling signal S, and corresponding output signal V_A' of a S/H circuit. The circuit samples the input on the rising edge of the switching signal S.

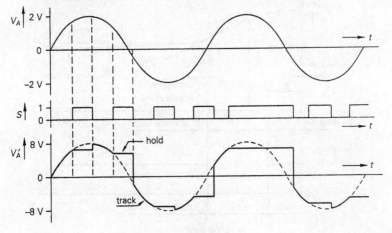

errors. After the 'hold' command is issued (S goes to '1'), there is another time delay, the aperture time, before the output signal becomes constant. The length of this time delay is not always constant. This causes an aperture time uncertainty. The actual sampling point lies within this uncertainty interval. The capacitor in the S/H circuit will not hold the sampled value exactly, but will charge or discharge slowly during the hold period. This is known as 'droop' or amplitude decay. Furthermore, the input signal will, to some minor degree, continue to affect the output voltage during the hold period. This is known as 'feed-through'.

Since the aperture time is usually small, the acquisition time and the conversion time of the ADC determine the maximum sampling speed. The aperture time causes a delay of the sampling point, which is a timing error. The aperture uncertainty causes an uncertainty of the actual sampling point. This is known as jitter. The transfer uncertainty determines how large the error in the sampled value V'_A becomes (for a constant input signal, when the aperture time does not contribute to the error). For extremely long hold times we also encounter a time dependent amplitude error due to the decay of the voltage across the holding capacitor.

We shall now discuss the theory underlying the sampling of a signal. In Fig. 4.20 the sampling points are not spaced evenly along the time axis. We can then no longer reconstruct the shape of the input signal. For certain

Figure 4.21. Acquisition and aperture time in an S/H circuit. NB The time intervals as well as the slope of the input signal V_A are greatly exaggerated for purposes of clarity.

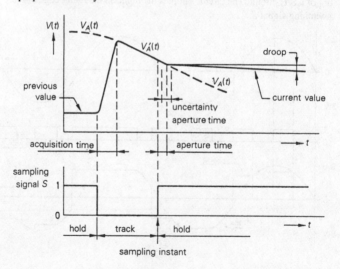

applications the sampling points are chosen at random along the time axis. With such random sampling the information about the waveform is lost. We can only still determine the amplitude probability density function from the random samples. Random sampling therefore gives us statistical amplitude information about the input signal. This means that we can still measure the RMS-value, peak value, range, etc. of the input signal, but not its shape or frequency content.

In most cases a signal is sampled at equispaced sampling points. An important issue is then how many samples are required per unit of time in order still to be able to describe reliably the time-continuous input signal. We do not wish to lose information in the sampling process, yet we do not want to sample too often either. The answer to this question lies in Shannon's sampling theorem. This theorem states that, in order to reconstruct (without error) the original input signal from equispaced samples of that signal, the sampling frequency f_s must be higher than twice the highest frequency f_{max} that is present in the continuous input signal. We must note here that 'input signal', as stated above, not only denotes the actual signal but also contains the distortion components and noise in the signal. The theorem assumes, therefore, that a maximum frequency f_{max} exists, above which the spectral power in the input signal is zero.

In order to see what happens when this requirement is not met, we shall determine the frequency spectrum that is created by sampling a time continuous signal. For simplicity, we shall only examine the value of the sampled signal during the sampling instants (see Fig. 4.22).

The sampled signal can be described as the product of the analogue input signal $V_A(t)$ and a signal $S(t)$ consisting of a series of equispaced unit impulses $\delta(t - mT_s)$. Therefore we have:

$$V'_A(t) = V_A(t)\, S(t)$$

with:

$$S(t) = \sum_{m=-\infty}^{\infty} \delta(t - mT_s)$$

where $\delta(t - mT_s)$ are so-called Dirac impulses. These are impulses which are non-zero only for $t = mT_s$ and which, when integrated with respect to time, yield an area of one (second). Since $S(t)$ is periodic in time with period T_s, it may be expressed in a Fourier series:

$$S(t) = \sum_{n=-\infty}^{\infty} C_n\, e^{jn\omega_s t}$$

The Fourier coefficient C_n is given by:

$$C_n = \frac{1}{T_s} \int_t^{t+T_s} S(t)\, e^{-jn\omega_s t}\, dt$$

Since $S(t)$ (on the interval $(t, t + T_s)$) is only non-zero for the points mT_s, the following holds:

$$C_n = \frac{1}{T_s} \int_t^{t+T_s} \delta(t - mT_s)\, e^{-jn\omega_s t}\, dt = \frac{1\ \text{second}}{T_s}$$

where m is chosen such that $mT_s \in (t, t + T_s)$. Since:

$$V_A'(t) = V_A(t)\, S(t) = \frac{1}{T_s} \sum_{n=-\infty}^{\infty} e^{jn\omega_s t}\, V_A(t)$$

the Fourier transform of $V_A'(t)$ is:

$$F'(j\omega) = F\{V_A'(t)\}$$

$$= \frac{1}{T_s} F\left\{ \sum_{n=-\infty}^{\infty} e^{jn\omega_s t}\, V_A(t) \right\}$$

$$= \frac{1}{T_s} \sum_{n=-\infty}^{\infty} F\{e^{jn\omega_s t}\, V_A(t)\}$$

$$= \frac{1}{T_s} \sum_{n=-\infty}^{\infty} F(j\omega + jn\omega_s)$$

Figure 4.22. (a) Equispaced sampled signal consisting of Dirac impulses. (b) Multiplication model for a Dirac sampler.

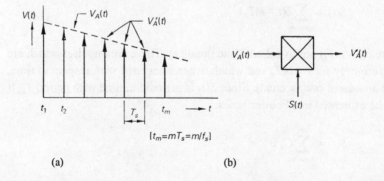

$[t_m = mT_s = m/f_s]$

(a)   (b)

Here $F(j\omega)$ is the Fourier transform of the input signal $V_A(t)$. The magnitude $|F(j\omega)|$ is identical to the amplitude frequency spectrum in V/Hz of the input signal $V_A(t)$. We notice from this calculation that the amplitude spectrum $F'(j\omega)$ of the sampled signal is equal to $1/T_s$ times the spectrum $F(j\omega)$ of the input signal, and that it repeats around the frequencies $n\omega_s$. As Fig. 4.23 shows, the spectrum of the sampled signal is therefore an infinite, equispaced repetition of the low-frequency input signal spectrum $F(j\omega)$ around the harmonics $n\omega_s$ of the sampling frequency ω_s. One such repetition already contains all the information about the input signal. A repetition is called an 'alias'. The zero-order alias lies in the base-band and the nth alias lies around the frequency $n\omega_s$.

This repetition also occurs in a less idealised sampling situation. If, for instance, the width of the sampling impulse is non-zero, and even if we hold the sampled values during the entire sampling period T_s with an S/H circuit, we still see this spectral repetition. However, the higher aliases will now have a lower amplitude than in the Dirac sampler case discussed above.

We see from Fig. 4.23 that, as long as the highest frequency f_{max} of the input signal remains smaller than half the sampling frequency f_s, we can recover the base-band (the zeroth order alias) from the spectrum $F'(j\omega)$ by means of a low-pass filter. Such a filter removes the higher aliases in the frequency domain. The filter reintroduces the missing parts of the sampled signal in the time domain. These filters are therefore called 'reconstruction'

Figure 4.23. Frequency spectra of the signals involved in the sampling in Fig. 4.22. $F(j\omega)$ is the spectrum of the input signal $V_A(t)$, $F_s(j\omega)$ is the spectrum of the sampling signal $S(t)$ and $F'(j\omega)$ is the spectrum of the sampled signal $V'_A(t)$.

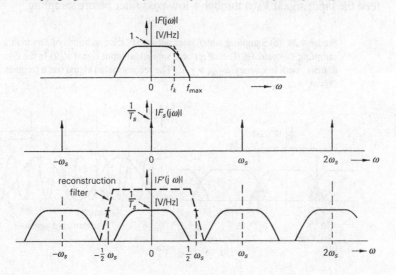

or 'interpolation' filters. If f_{max} were to come very close to the frequency $\frac{1}{2}f_s$ we would need a low-pass filter with a very steep roll-off slope (a brick wall filter). It is not possible to build 'real time' filters that have such a characteristic. We could, however, record the signal and afterwards, when the entire signal is known, use mathematical manipulations to reconstruct the input signal. We would then have created a 'non-causal' filter; this is a filter that uses knowledge about the future variations of the signal. The practical conclusion is therefore that, as long as $f_{max} \leq \frac{1}{2}f_s$, we can reconstruct the input signal without error.

This is no longer the case if this condition is not met, as illustrated in Fig. 4.24. We see that the aliases begin to overlap here. It is then no longer possible to reconstruct the zero order alias without error, even with a brick wall low-pass filter.

The results in the time domain are illustrated in Fig. 4.24(b). The sinusoidal input signal $V_A(t)$ with frequency ω_{max} is undersampled. We feed the sample pulses through a low-pass filter with cut-off frequency $\frac{1}{2}\omega_s$ in an attempt to reconstruct the input. This results in a sine wave with the same amplitude as the original, but with the frequency $\omega_s - \omega_{max}$. This is also evident from the spectrum of the sampled signal, shown in Fig. 4.24(a). The signal energy is in essence mirrored at the frequency $\frac{1}{2}\omega_s$. The errors that result from the shaded sections of the overlapping aliases are known as 'aliasing errors'. A common example of aliasing is when the spokes in the wheels of a carriage in a Western film appear to be rotating in reverse (the spoke frequency and film frame frequency do not meet Shannon's condition). From Fig. 4.24(a) we see that we can avoid these aliasing errors if we feed the input signal $V_A(t)$ through a low-pass filter before sampling.

Figure 4.24. (a) Sampling which does not satisfy the conditions of Shannon's sampling theorem. (b) The effect on a sinusoidal input signal $V_A(t)$ in the time domain, with frequency $\omega_{max} > \frac{1}{2}\omega_s$. The reconstructed signal has a frequency $\omega = \omega_s - \omega_{max}$.

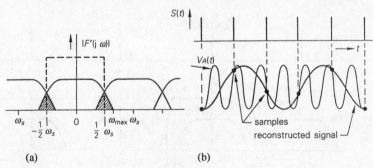

(a)   (b)

Such a filter, called a 'presampling' filter, reduces the high-frequency energy in the spectrum of $V_A(t)$, thereby reducing, or even eliminating, the aliasing errors. However, we also introduce an error, since, as the cut-off frequency of the presampling filter decreases, it removes an increasing portion of the signal $V_A(t)$. The conclusion is therefore that a prefilter must not be used in a DAS to reduce the effects of undersampling, but only to remove (the always present) high-frequency noise and distortion components from $V_A(t)$, while sampling at a correctly chosen rate. As is illustrated in Fig. 4.25, if these unwanted components were left in, they would mirror or 'fold back' and appear in the high-frequency portion of the spectrum of $V_A(t)$. For this reason the input stage of a DAS almost always contains a low-pass filter. The other filter in a DAS is located at the very end of the signal conversion chain. This is the reconstruction filter which interpolates the DAC output signal between the sampling points. In the following section we shall discuss the reconstruction process.

4.5.5 Reconstruction theory

In the previous sections we have seen that sampling an analogue signal $V_A(t)$ results in a time-discrete signal. This is necessary since a digital processor can only receive and output information at discrete points in time; a DAS requires a sampler at the input side. At the output side this time-discrete signal must be converted to a time-continuous signal again in order to arrive at an analogue output signal. This conversion is performed by a 'reconstruction' or 'interpolation' filter. In the time domain, sampling makes an analogue (or time-continuous) signal time-discrete, while reconstruction

Figure 4.25. (a) Spectrum $F(j\omega)$ of a signal with noise and distortion, and the associated spectrum $F'(j\omega)$ of the sampled signal. (b) Spectrum $F(j\omega)$ of the input signal after the 'pre-sampling' filter and the corresponding spectrum $F'(j\omega)$ of the sampled signal.

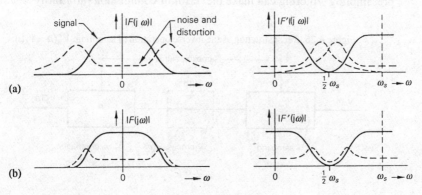

makes this time-discrete signal time-continuous again. In the frequency domain, sampling creates higher order (or secondary) aliases of the spectrum of the input signal, while the reconstruction removes these higher-order spectra in order to recover only the base-band (or primary) spectrum. Since we focus here on (avoiding) errors and inaccuracies in DASs, we may conveniently ignore the useful operations performed on the digitised signal by the processor. We therefore assume that the digital processor has a unity transfer function: i..e. we pretend there is no processor in the DAS. We then arrive at the model of a DAS illustrated in Fig. 4.26.

The analogue input signal $V_A(t)$ is sampled in S. The sampled signal $V'_A(t)$ is reconstructed in R. If we perform ideal sampling and reconstruction, the output of the reconstruction yields an exact replica of the input signal: $V''_A(t)$ = $V_A(t)$. Ideal reconstruction is therefore an operation which is the exact inverse of ideal sampling! In general though, errors are introduced by both operations. The errors which result from sampling were discussed in the previous section. We saw that, in order to avoid aliasing errors, a 'presampling' filter is also used. In practice, this filter will always remove a fraction of the useful signal, thereby creating 'errors of omission'. Also, during the reconstruction we remove the trailing edge of the base-band spectrum. This also contributes to the omission error.

In Fig. 4.24(a) the omission error is the portion of the base-band spectrum that extends outside the interval $(-\frac{1}{2}\omega_s, +\frac{1}{2}\omega_s)$; it is filtered out. A DAS also makes 'errors of commission'. These result from its inability to filter out the first alias in the reconstruction stage completely. In Fig. 4.24(a) this error arises from the section of the first alias that lies in the interval $(-\frac{1}{2}\omega_s,$ $+\frac{1}{2}\omega_s)$. The errors of omission and commission are shaded in this figure. The sum total of these errors makes up the overall 'aliasing error'. We may now state that in order to reconstruct $V_A(t)$ appropriately from the sampled signal, we must reduce the omission and commission errors as much as possible and preferably to zero. From the above it will be clear that excessive 'presampling' filtering can make the 'error of commission' arbitrarily small.

Figure 4.26. Reconstruction as the inverse operation of sampling: $V''_A(t) = V_A(t)$.

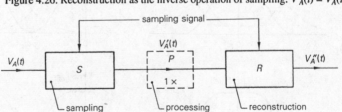

The 'error or omission', however, will then grow very large. Conversely, by applying no prefiltering and no reconstruction filtering at all, the error of omission may be reduced to zero, but the error of commission will be large. One therefore expects that the total aliasing error can be minimised if the proper balance is struck between omission and commission errors. We may further expect that this minimum will decrease as the sampling frequency is increased. In order to examine these errors more closely, we shall first discuss a number of reconstruction methods in greater detail.

At the output side of the DAS, the DAC will give us a step-wise approximation of the desired analogue output signal, as is illustrated in Fig. 4.27(a). At the instants when the digital input signal to the DAC changes, unwanted voltage spikes may result. For this reason, one usually employs an S/H circuit behind the DAC that samples only when the output of the DAC has settled down. This results in a so-called 'deglitched' DAC.

One can visualise the step-wise signal created by a deglitched DAC as a signal generated from the input signal taken at the sampling times mT_s, which are held constant during a time T_s. Such a DAC is therefore referred to as a 'zero order hold' circuit. This circuit fills in the (analogue) output voltage between the sampling points, and can thus be thought of as a reconstruction or interpolation filter. It may be shown that the transfer function $H(j\omega)$ of this filter is:

$$H_0(j\omega) = \frac{\sin(\pi\omega/\omega_s)}{\omega/\omega_s} \, e^{-j\pi\omega/\omega_s}$$

Here ω_s is the sampling frequency. The transfer function of such a zero order hold filter is calculated using Fig. 4.26, replacing S with an ideal Dirac sampler and R with a zero order hold circuit. The frequency response of this reconstruction filter is shown in Fig. 4.28. One can see that the filter introduces no phase distortion; the phase shift linearly increases with frequency.

Figure 4.27. (a) Zero order hold circuit as a reconstruction filter in the time domain. (b) Predictive first order hold circuit (linear extrapolation). (c) Interpolating first order hold circuit (linear interpolation). NB The time delay of one sampling period that results from this method is not shown here.

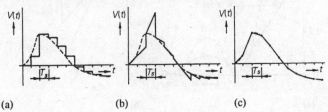

(a)   (b)   (c)

We can calculate the maximum instantaneous error produced by such a circuit for a sinusoidal signal $V(t) = A \sin \omega t$. This maximum error occurs at the zero crossings. There, the slope of the signal is given by $dV(t)/dt = A\omega$. This is simply equal to the maximum error ΔV, divided by the sampling period T_s. Therefore, we can write: $\Delta V = A\omega T_s$. The maximum relative amplitude error then becomes:

$$\varepsilon_0 = \omega T_s = \frac{2\pi\omega}{\omega_s} \quad (\omega T_s \ll 1)$$

We must therefore sample a sinusoidal waveform at least 628 times per period in order to arrive at a maximum amplitude error of less than 1% if one uses a zero order hold circuit for analogue signal reconstruction. Theoretically, however, we have seen that a sampling frequency of $\omega_s > 2\omega$ will be sufficient for an error free reconstruction. This large discrepancy in required sampling frequencies stems from the fact that a zero order hold circuit is easily implemented, yet does not make an effective reconstruction filter.

Fig. 4.27(b) illustrates the time-domain response of a so-called 'predictive first order hold' circuit. This circuit extrapolates the slope of the signal during the last sampling interval, and assumes the value of the input signal at the sampling points. It is therefore also known as a 'linear extrapolator'. When this is used as a reconstruction filter, the transfer function of a linear extrapolator is given by:

$$H_1(j\omega) = T_s^2(1 + j\omega T_s)\left(\frac{1 - e^{-j\omega T_s}}{j\omega}\right)^2$$

Figure 4.28. (a) Amplitude characteristics of a zero order hold circuit and a first order extrapolating hold circuit, both used as a reconstruction filter. (b) The corresponding phase characteristics.

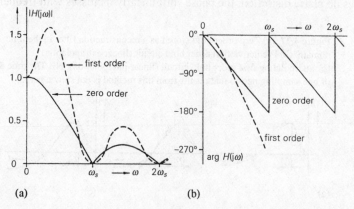

(a)   (b)

This transfer function is illustrated in Fig. 4.28. One may notice that this method of reconstruction actually amplifies the higher frequencies, and does not have a linear phase response. Due to the large phase shift for high frequencies, there is a distinct possibility for dynamic instability with a DAS employing this type of reconstruction filter as part of a feed-back loop. The stability of a low-pass system is usually compromised by the high-frequency range of the transfer. We can calculate that the maximum instantaneous amplitude error for a sinusoidal signal is given by:

$$\varepsilon_1 = (\omega T_s)^2 = \frac{4\pi^2 \omega^2}{\omega_s^2} \quad (\omega T_s \ll 1)$$

In a first (and also in a higher) order hold circuit the maximum error occurs around the peaks of a sinusoidal signal. For an error of 1% we still need approximately 63 samples per period (sinusoidal signal).

Fig. 4.27(c) shows the signals in the time domain for another category of reconstruction circuits: the linearly interpolating first order hold circuits. Such a circuit interpolates linearly between sampling points and, at the sampling points, this circuit provides the sampled input values. Of course, this is only possible if both boundary values of the interpolation interval are known. The output signal of an interpolating reconstruction circuit therefore will appear with a delay of at least one sampling period, with respect to the input signal. We can calculate that a linearly interpolating reconstruction filter exhibits a maximum instantaneous amplitude error for a sinusoidal input given by:

$$\varepsilon_{11} = \tfrac{1}{4}(\omega T_s)^2 = \frac{\pi^2 \omega^2}{\omega_s^2} \quad (\omega T_s \ll 1)$$

This maximum error again occurs around the peaks of the signal. The maximum instantaneous amplitude error is now smaller than 1% for only 16 samples per period, again assuming a sinusoidal input signal.

We can also implement a higher order interpolation. This is usually computed inside the processor of the DAS. To this end the processor must be much faster than the sampling rate at the input side of the DAS. The processor will calculate n intermediate values per DAS input sampling interval. These are then output via a DAC and, for instance, a zero order hold circuit running at an output frequency n times greater than the input sampling frequency. Even if the processor is not fast enough, or if, due to system architectural reasons, it is not possible to perform this higher order interpolation in the processor itself, it is still possible to perform digital interpolation. This is then accomplished by means of a separate 'post processor'. Since the overall error of a DAS is commonly dominated by the reconstruction error, and since the higher order interpolation discussed above is a form of digital filtering not affected by parameter drift, it is usually beneficial to perform such digital postprocessing. It should be mentioned that analogue electronic reconstruction filters, however, are affected by parameter drift. In addition, the digital postprocessing filters are locked in to the changes in the sampling frequency, and therefore do not need to be tuned like their analogue counterparts.

One may employ an analogue filter (such as a Butterworth filter) as a reconstruction filter. If the sampling theorem criterion is met (if $\omega_s > 2\omega_{max}$), it should be possible to reconstruct ideally (without error) the input signal using a low-pass filter that has a flat amplitude characteristic and a linear phase characteristic up to $\frac{1}{2}\omega_s$, provided that it has an infinitely sharp roll-off at $\frac{1}{2}\omega_s$ (brick wall filter). As we saw previously, however, such a filter cannot be implemented in real time. We may recall that this would lead to a so-called non-causal filter. A filter with an infinitely sharp roll-off would contradict the causality principle since the step response of such a signal starts even before the step is applied at the input. These filters can therefore only be implemented 'off line'. The signal is first recorded, or sent into a delay line, after which one determines not only the value at instant t_i, but also the values preceding t_i and the values the signal will assume in the future. Therefore such filtering schemes are usually only implementable in the form of a digital filter algorithm in a computer or digital processor.

We have seen that aliasing errors depend on the wave shape and the frequency content of a signal. They are determined especially by the spectral power of the signal around $\frac{1}{2}\omega_s$. Since measurement instrumentation deals with a multitude of different types of signal, it is not feasible to perform an error calculation for each of these (usually not known *a priori*) signals. It is therefore common to use a noise signal as a test signal to characterise the accuracy of a DAS.

'White noise' (i.e. frequency independent noise) with a normal amplitude probability density (Gaussian amplitude distribution) is applied as input to the system. This noise is passed through a so-called 'shaping filter'. This filter is a low-pass filter which is flat up to the roll-off frequency ω_b, after which it rolls off with $6n$ dB/octave (giving it a spectral roll-off slope of $(\omega_b/\omega)^n$). The noise at the output of this nth order filter is usually designated 'nth order data' with a bandwidth of ω_b. The RMS-value of this input signal is then determined (standard deviation of the Gaussian amplitude distribution). This test signal is passed through a DAS with a given sampling frequency and a given antialiasing and reconstruction filter. The RMS-value of the reconstructed signal is then measured. From this we can determine the error of the DAS and its dependence on the sampling frequency as well as of the type and order of the antialiasing and reconstruction filters used. In this characterisation, the error is actually determined on a signal power basis and not, as shown above for a sinusoidal signal, on the basis of the maximum instantaneous amplitude error. Some results are illustrated in Fig. 4.29 for second order ($n = 2$) data and for fourth order ($n = 4$) data.

The relative error ε_t is plotted here as a function of the ratio (ω_s/ω_b) of the sampling frequency ω_s and the bandwidth ω_b of the DAS input (noise) data spectrum. ε_t is the total error in the RMS-value of the reconstructed data signal. In this figure we assume only reconstruction errors; no antialiasing filter is present. Based on the signal power present in the high-frequency end of the data spectrum, we may expect that for a reconstruction filter with a sufficiently sharp roll-off the error is:

$$\varepsilon_t = \left(\frac{\omega_s}{\omega_b}\right)^{-(2n-1)/2} + C$$

where n is the order of the data and C is an arbitrary constant.

From the above we see that reconstruction errors can easily lead to large errors in a DAS. Let us now also take into account (anti)aliasing as a source of errors. If we are free to choose the roll-off frequency of the reconstruction filter (which is the case with analogue filters) we can optimise the DAS with respect to the total error ε_t. Fig. 4.30 shows the result of this optimisation between errors of omission and errors of commission. Here, for a specific sampling frequency $\omega_s = 10\omega_b$, we vary the roll-off frequency ω_f of the

Figure 4.29. Reconstruction error for nth order data in different reconstruction filters. A: Zero order hold circuit. B: Fourth order Butterworth filter. C: Linearly interpolating first order hold circuit. (a) $n = 2$. (b) $n = 4$.

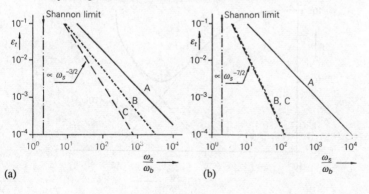

reconstruction filter from $\omega_f = 0.1\omega_b$ to $\omega_f = 10\omega_b$. Fig. 4.30 illustrates the situation for a fourth order Butterworth reconstruction filter and second order data with a bandwidth of ω_b.

The total error ε_t is made up of an error of omission ε_o and an error of commission ε_c such that:

$$\varepsilon_t = \sqrt{\varepsilon_o^2 + \varepsilon_c^2}$$

For a low value of ω_f we filter away too much of the zero order alias (error of omission dominates) and for a high ω_f we pass through too much of the first order alias (error of commission dominates). The total error ε_t is minimised at the point where $\omega_f \approx 3\omega_b$. In the curves of Fig. 4.29 these minimum values of ε_t are illustrated for various reconstruction filters and data orders.

4.5.6 Multiplexing

In the previous sections, we talked only about a single-channel DAS: i.e. one in which there was only one input signal to be measured, and one output control signal to be reconstructed. In practice, it is important to use the signal conversion and reconstruction sections, as well as the processor, more efficiently. This is accomplished by so-called 'multiplexers' (signal scanners or commutators) which read all the input signals into the computer by toggling through each signal in rapid succession. This is a form of time multiplexing (see Section 2.2). All input signals then use the same processor, conversion and distribution sections. These are used in a time-sharing fashion, yielding a multichannel DAS. One advantage of such a

Figure 4.30. The total error ε_t, the 'error of omission' ε_o and the 'error of commission' ε_c for a fourth order Butterworth filter and second order data.

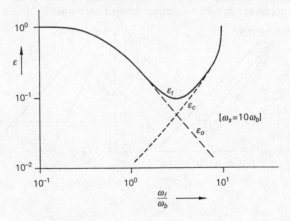

system is that the cost per channel (and thus per input signal) is much lower than those for a single-channel DAS. A multichannel DAS usually contains 10–100 channels, but systems featuring up to 1000 channels are in industrial use.

The cost advantage, however, can be offset by the problem that multiplexers (and demultiplexers at the output) introduce extra transfer errors and that the effective sampling rate is reduced as the number of channels increases. In a k-channel DAS, the highest frequency of each of the k input signals may not exceed $1/k$ times the frequency allowed in a single-channel DAS using the same conversion circuits. The demultiplexer at the output of a multichannel DAS simply performs the inverse function of the multiplexer, and will therefore not be discussed separately in this text.

A multiplexer is usually made up of electronic switches. In the past, mechanical switches (cross-bar scanners) were used, but currently all switching is done electronically. In the simplest case a multiplexer is made up of k switches which connect one out of k inputs to a single output. To prevent shorting two inputs together, these switches are of the 'break-before-make' type. A simple multiplexer only switches when it receives a command from the DAS 'timing unit'. The k channels of this so-called 'sequential multiplexer' are then switched to the output in the order in which they are connected at the input. A more flexible solution contains a 'random access' multiplexer, in which a central computer determines which channel is to be connected to the output by means of a channel address. A random access multiplexer allows for so-called 'adaptive multiplexing', a scheme by which the low-bandwidth input signals are addressed less frequently, and thus sampled at a lower rate than the high-bandwidth input signals. This method allows more economical use of the conversion and processing speed of a DAS when the input signals differ greatly in frequency spectrum.

Three forms of commutation are possible in a multiplexer: commutation, supercommutation, and subcommutation. As shown in Fig. 4.31, commutation is implemented by a multiplexer which is built as described above; one (effective) switch connects one out of k inputs to the output. A multiplexer in which a single input signal is connected to multiple input ports utilises 'supercommutation'. In supercommutation, the number of input ports to which a signal is connected determines the effective sampling rate of that signal. This method can therefore be used as a hard-wired solution for dealing with wide-band input signals. It is obvious that this is accomplished at the expense of the number of available input channels. In the example of Fig. 4.31(b) the first two channels have a sampling rate which is four times greater than that of the others; the number of available channels is reduced here by six.

Direct commutation, as shown in Fig. 4.31(a), is not easy to implement practically, especially in large multiplexers. The problem lies in cross-talk and feed-through to the output from the input signals connected to the channels which are switched off. The problem occurs because the (field effect) transistors used in the switches are not ideal. Additionally, each switch introduces a parasitic (load) impedance between the input and ground. The noise levels and the loading effects would be intolerable if we placed large numbers of these switches in parallel. Therefore, large multiplexers use 'subcommutation'. As shown in Fig. 4.31(c), subcommutation is implemented using multiple stages. In the example shown in the figure, a single channel in the multiplexer of Fig. 4.31(a) would suffer from cross-talk from say 15 other channels, while the one shown in Fig. 4.31(c) would limit the crosstalk to only 3 channels and 3 subchannels.

It is especially important to guard against the introduction of noise when using multiplexers for small signals in a noisy environment. Fig. 4.32(a) indicates how large signals are usually measured. Even though the signal source V_v is located some distance away from amplifier A, the voltage V_g

Figure 4.31. (a) Commutation of k channels. (b) Supercommutation of channels 1 and 2. (c) Subcommutation of k channels into 4.

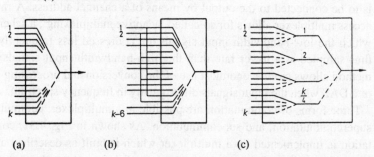

(a)   (b)   (c)

Figure 4.32. (a) Asymmetrical multiplexer. (b) Symmetrical multiplexer. (c) Pseudo-symmetrical multiplexer. V_g is a noise voltage between the grounds of A and V_v due to stray ground currents.

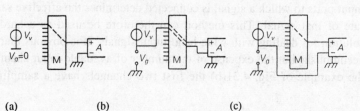

(a)   (b)   (c)

between the ground points of A and V_v is negligible compared to V_v. A multiplexer structure which is asymmetrical (with respect to ground) is then sufficient. Such an 'asymmetric multiplexer' is usable if $V_g = 0$ or if V_g/V_v is very small. If these requirements are not met, then a 'symmetrical multiplexer' (as shown in Fig. 4.32(b)) must be used.

In this multiplexer, two switches are used to switch one single input channel. The 'common-mode rejection' of the amplifier A is then also important (see Section 2.3.3.3). A simpler, but also somewhat less effective solution, is illustrated in Fig. 4.32(c). In using the 'pseudo-symmetrical multiplexer' shown here we assume that all input signal sources have approximately the same noise voltage V_g. This is usually only true as an approximation, however; this method of multiplexing does not offer the level of common-mode noise immunity that a symmetric multiplexer would yield. In order to measure small signals precisely, shielding (as discussed in Section 2.3.3.3) must also be used. There are multiplexers in which the shield is also switched; one therefore needs three switches per channel.

The central digital processor in a DAS can only take in information at discrete points in time. If the processor needs to perform an algorithm on multiple input signals, we usually assume that these signals were all sampled at the same instant in time. In practice, however, with the multiplexers discussed above, the signals are sampled in sequence. Corresponding samples are therefore shifted in time by some small amount. This time delay between the samples of a number of parallel signals is known as 'skewing'. The skewing causes an unwanted phase shift in the processing. It is possible to minimise the skewing by connecting the signals that are to be processed in one algorithm in such a way that they all enter adjacent input channels of the multiplexer. This way they are sampled only a short time apart. If this method still introduces too much phase shift, one can use a so-called 'simultaneous multiplexer'. This may be more expensive, as indicated in Fig. 4.33, but it is the only correct solution to avoid skewing altogether.

As we see in Fig. 4.33(a), every input of the multiplexer (M) is fed through a sample and hold circuit (S/H). A timing unit toggles all these circuits jointly into the 'hold' mode. This is done once per scan cycle of the multiplexer, at the frequency ω_s/k. The multiplexer then reads in the various 'frozen' input voltages, and connects them to an ADC. It is important that, in order to achieve perfect simultaneous sampling, the 'presampling' filters of each of the channels be identical. If they are not, undesired phase shifts will be introduced. For comparison, Fig. 4.33(b) depicts a conventional multiplexer, which uses only a single S/H circuit.

A multiplexer has a certain maximum useful scan rate. This is determined by the maximum number of channels that can be fed through to the output per second. The scan speed limitation is caused by the finite amount of time necessary to allow the switching transients, which occur when switching to a new channel, to decay. This time, which must pass before an output value is reached within the allowable tolerances, is known as the 'settling time'. The maximum scan rate is also known as 'throughput' (given in channels per second). Additionally, the multiplexer is not an ideal component. It has a non-zero resistance in the 'on' position, produces a small zero-offset error, and does not isolate the switched-off channels completely. Even in the switched-off state there is a certain amount of feed-through of an input signal to the output of the multiplexer. A careful selection of the multiplexer for a given application, however, will usually allow these errors to remain small.

4.5.7 Instrumentation computer systems

An instrumentation computer system is made up of a (mini)computer for 'real time' or 'on-line' processing. This computer is equipped with a data acquisition system for the measurement of physical signals, and a data distribution system for controlling of physical processes. The components for such a measurement and control system have been discussed in the previous sections. In this section we shall examine the complete system.

In a global sense, one can design the instrumentation computer system in two ways: centralised, or decentralised. Fig. 4.34 gives an example of a 'centralised' architecture. We call this system centralised because the signal conversion sections are used by all the signals sequentially. The associated circuitry is therefore usually located at the central computer. The advantages of this system are clear: through time sharing of the conversion sections, the

Figure 4.33. (a) Simultaneous multiplexer. (b) Conventional multiplexer.

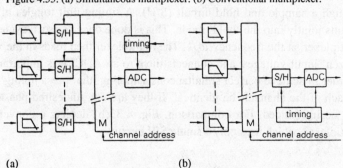

(a)   (b)

system cost is low. Fig. 4.35 shows an instrumentation system with a 'decentralised' architecture. In this architecture, every channel contains its own conversion sections; only the digital processor is used in a time multiplexing fashion. This method allows each channel to be optimised individually. In addition, the conversion sections used in this architecture may also be k times slower than those in the centralised scheme. These separate conversion sections will therefore be less expensive. In this method, the conversion can be done locally, at the location of the signal source, which means that digital (instead of analogue signals, which are very sensitive to noise) can be sent from the measurement source to the processor. Furthermore, each channel can be equipped with its own preprocessor, offloading the main processor, through the use of microcomputers. The processor in an instrumentation system can be connected to other processors using a 'bus' connection. This architecture is referred to as a distributed instrumentation system. The processor can also be connected with autonomous measurement systems outside the DAS. This is usually done through a standardised bus (the IEEE-488 bus, for example). Fig. 4.36 gives an example of such a system.

If the channels in a centralised DAS differ greatly in input signal amplitude, one can equip the central portion of the DAS with a program-

Figure 4.34. Centralised instrumentation system.

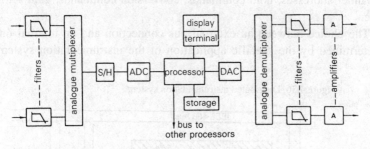

Figure 4.35. Decentralised instrumentation system.

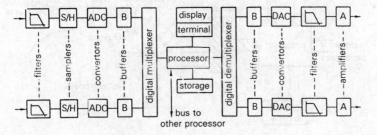

mable amplifier. The gain of this amplifier is then varied by the computer, simultaneously with the channel address. Setting this amplifier to a new gain, however, takes a certain amount of time, and reduces the maximum scanning rate of the system. It is therefore important to separate the channels into groups which have similar signal amplitudes. This prevents the programmable amplifier from having to switch every time a new channel is addressed. An even better method is the use of subcommutation. One can then choose three input multiplexers (a low, a medium, and a high-level multiplexer), for example, which are followed by amplifiers with a fixed gain, after which the signals are multiplexed into a single channel.

A 'reference' or 'calibration' signal is often connected to one of the system input channels. Transfer deviations may then be noted, and the digital processor may be programmed to correct for these errors.

A DAS must often also read in a number of binary inputs, such as the position of switches, or outputs from detection systems (for fire, overload, or other undesirable events). In Fig. 4.35, these signals are applied directly to the digital multiplexer as separate inputs. In a system as shown in Fig. 4.34, these signals are input to the processor either separately or via a digital multiplexer which switches between the ADC output and the digital multiplexer. Finally, a DAS must also contain a 'timing and control interface' which is controlled by the computer. This interface provides the drive signals which are necessary for housekeeping inside the DAS, generating channel addresses, hold commands, conversion commands, gain settings, etc.

The choice between an external bus connection and an internal one is determined by the specific application of the instrumentation system. A

Figure 4.36. Distributed instrumentation system.

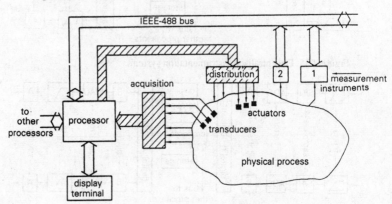

large project, for instance, the instrumentation of an oil refinery, will emphasise internal bus connections which uniquely fit the specific application. If one needs an instrumentation system, however, that is to be used only once, for example, to watch an expensive machine during its initial start-up, the emphasis will be placed on using a standardised external bus, to which readily available measurement devices may be connected. In the measurement and instrumentation world, one almost exclusively employs the IEEE-488 bus (known under several different names) for this purpose.

IEEE-488 bus

The IEEE-488 bus was developed in 1972 by Hewlett-Packard (HP), who gave it the name HPIB (Hewlett-Packard Interface Bus). This bus is also known as the GPIB-bus (General Purpose Interface Bus). In 1976 this bus was standardised by the European International Electrotechnical Commission as the IEC 625-1 bus. In 1978 it was standardised by the American Institute for Electrical and Electronics Engineers as the IEEE-488 bus. This international standardisation had the result that currently, almost every respectable measurement instrument is equipped with an IEEE-488 bus. This bus provides a flexible interface between measurement instruments, computers, and peripherals (such as plotters, printers, etc.).

The IEEE-bus is a so-called 'party line bus'; all participating devices are connected in parallel to the bus. The bus, comprising 16 conductors or lines, therefore runs along every participant. The bus allows a maximum of 15 hook-ups. In principle, every participant can communicate with every other device (bi-directionally) and transmit measurement or instrument control data. Every participant must be able to perform at least one of the three following functions: 'talker', 'listener', or 'controller'. A 'talker' transmits data to other participants via the bus, a 'listener' receives data via the bus. Many instruments are capable of being both talker and listener; a programmable instrument receives its control data in the listener-role and transmits its data in the talker-role. The simplest system comprises one talker connected to one listener. The data are then transmitted in one direction. The talker is manually placed in the 'talk only' mode, while the listener is placed in 'listen only' mode. One can thus connect a measurement device to, for instance, a plotter or chart recorder.

In the above, the 'controller' is a bus manager that indicates which device is the talker and which are the listeners. The controller can also place one or more devices in a different instrument mode, so that the device performs a different measurement function.

The 16 lines of an IEEE-bus carry Transistor–Transistor Logic (TTL)-level 'open collector' signals. The lines are subdivided into three subbuses, each with its own specialised function:

a. Data input/output (DIO) bus: This bus is made up of eight lines which carry data in a bit-parallel and a byte-serial format. The data transport along the DIO bus is asynchronous and uses a so-called 'handshake' procedure between the participants.
b. Handshake bus: The handshake bus is made up of three lines which together handle the handshake for data transport along the DIO bus. Only when the slowest participant is finished reading and accepting the data is the bus released for the next phase. The disadvantage of using this method is that, if one of the participants fails to post the 'ready' flag the bus stays blocked from further use. The controller must then, after a 'time-out' period, restore the bus to normal use.
c. Management bus. This bus is made up of five lines which are used to ensure the orderly transmission of messages along the bus.

We shall now briefly explain the working of each of these three subbuses. Let us begin with the data bus. This bus does not carry only data (measured data, control information) but also participant addresses, universal commands, and bus-status bytes to and from the participants. The type of data offered on the DIO lines is determined by the ATtentioN line (ATN) of the management bus. If the ATN bit is 'true', there are addresses or universal commands on the data bus, to which all participants must listen. If the ATN bit is 'false' the data bus carries data designated only for those devices which were already assigned as talkers and listeners. It is important to note that all the lines of the handshake and management buses use negative logic; a 'true' (logical one) is a low (TTL) voltage, and 'false' (logical zero) is a high (TTL) voltage.

The handshake bus is made up of three predestined signal lines: 'DAta Valid' (DAV), 'Not Ready For Data' (NRFD) and 'Not Data ACcepted' (NDAC). If the DAV line is asserted (low) the talker is indicating that the data on the DIO lines is valid. The talker cannot change this data until the NDAC line is deasserted (high). This is after all listeners have read and accepted the data. When the NRFD line is asserted (low) the listeners are indicating that:

1. they all see correct data on the DIO lines (DAV asserted); and
2. they are all ready to read these data.

The NDAC line is asserted (low) during the handshake if not all listeners have read and accepted the data; the talker may not yet change or remove the data from the DIO lines. Fig. 4.37 shows a timing diagram of a handshake example. We see that the handshake procedure is 'locked'; it waits until the slowest instrument is ready for the next phase. The three signals NRFD, DAV, and NDAC are synchronised as indicated with the arrows in Fig. 4.37. The two lines NRFD and NDAC are used in the form of a 'wired OR' so the participants can interrogate every other device on the bus. This is shown in Fig. 4.38. This method of switching ensures that the system waits

Figure 4.37. Timing diagram of the locked handshake procedure that accompanies the data transport over an IEEE bus. During time interval *A*, a talker puts data on the DIO bus. Time interval *B* is used to read in this data. During time interval *C* the read in command is processed in the participant.

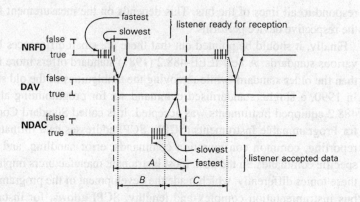

Figure 4.38. The two lines NRFD and NDAC of the handshake bus are implemented in each of the participants as a 'wired OR'. If one or more participants are not yet ready to read the data, the switch is closed. The NRFD line of the IEEE bus is then asserted (low) despite the fact that other participants may be ready (switch opened). If a participant is the talker, the deassertion of the NRFD is used to give the DAV signal.

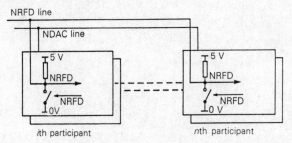

until the slowest instrument is ready before giving the flags 'Ready For Data' and 'Data ACcepted' are set by deasserting the NRFD and NDAC lines.

The management bus has one line called 'InterFace Clear (IFC)' which is asserted (low) by the controller in order to bring the entire IEEE interface to a desired initial state. A second line called 'Service ReQuest (SRQ)' may be asserted (low) by a participant if it needs attention or needs to interrupt the current order of processing. The 'Remote ENable (REN)' line is used to select between two alternative sources of control data. The 'End Or Identify (EOI)' line is used to mark the end of a series of data bytes, or to execute, in conjunction with the ATN line, the order of transmission of the participants. The function of the ATN line was discussed earlier.

In this text we can only present the general idea of an IEEE bus. The details depend on the specific participants used. Not all devices need to respond to all lines of the bus. This depends on the measurement function the respective device performs.

Finally, it should be pointed out that there are small differences between various standards. A new IEEE-488.2 (1987) standard offers more freedom than the older standard, while removing the ambiguities of the old standard. In 1990, a single, standardised command set for programming all IEEE-488.2 equipped instruments was accepted. It is called Standard Commands for Programmable Instruments (SCPI). SCPI addresses data formats, status reporting, common configuration commands, error handling, and device-specific commands. In the past, most instrument manufacturers implemented these topics differently, which made the development of the programming of bus instrumentation complex and lengthy. SCPI allows, for instance, all voltage measurement equipment to perform a voltage measurement with the simple command: MEAS:VOLT?

Appendix

In this appendix, several related items will be discussed, which provide further explanation or additional information on the subjects which have already been covered in the previous chapters.

A.1 SI units

The international system of metric units (Système International d'Unités), as accepted during a number of meetings of the Conférence Générale des Poids et Mesures, consists of seven base quantities, two additional quantities and numerous derived quantities. The seven base quantities are given with their SI units in Table A.1, the two additional units are given in Table A.2 and several examples of derived quantities with their SI units are provided in Table A.3. In addition, Table A.4 lists the decimal prefixes, used for indicating multiples and submultiplies of a thousand.

Table A.1. *Base quantities and units of the SI system.*

Base quantity	SI unit	
	Name	Symbol
length	metre	m
mass	kilogram	kg
time	second	s
electrical current	ampere	A
thermodynamic temperature	kelvin	K
luminous intensity	candela	cd
amount of substance	mole	mol

Table A.2. *Additional quantities and SI units.*

Quantity	Unit	Symbol
(plane) angle	radian	rad
solid angle	steradian	sr

315

The base units are defined a follows:

The SI unit of measure for the base qunatity *length* is the *metre*. One metre is the distance travelled by light in a vacuum during a time interval of 1/299 792 458 s.

Table A.3. *Derived quantities with their own SI units.*

Quantity	Unit	Symbol	Derivation		
frequency	hertz	Hz			s^{-1}
force	newton	N			$kg \cdot m/s^2$
pressure (stress)	pascal	Pa	N/m^2	=	$kg/m \cdot s^2$
energy, work	joule	J	$N \cdot m$	=	$kg \cdot m^2/s^2$
power	watt	W	J/s	=	$kg \cdot m^2/s^3$
electric charge	coulomb	C			$A \cdot s$
electric potential	volt	V	W/A	=	$kg \cdot m^2/s^3 \cdot A$
electric resistance	ohm	Ω	V/A	=	$kg \cdot m^2/s^3 \cdot A^2$
conductance	siemens	S	Ω^{-1}	=	$s^3 \cdot A^2/kg \cdot m^2$
capacitance	farad	F	C/V	=	$s^4 \cdot A^2/kg \cdot m^2$
inductance	henry	H	Wb/A	=	$kg \cdot m^2/s^2 \cdot A^2$
magnetic flux	weber	Wb	$V \cdot s$	=	$kg \cdot m^2/s^2 \cdot A$
magnetic flux density	tesla	T	Wb/m^2	=	$kg/s^2 \cdot A$
luminous flux	lumen	lm			$cd \cdot sr$
illuminance	lux	lx			$cd \cdot sr/m^2$

Table A.4 *Decimal prefixes.*

Prefix	Abbreviation	Factor
yotta	Y	10^{24}
zeta	Z	10^{21}
exa	E	10^{18}
penta	P	10^{15}
tera	T	10^{12}
giga	G	10^{9}
mega	M	10^{6}
kilo	k	10^{3}
milli	m	10^{-3}
micro	μ	10^{-6}
nano	n	10^{-9}
pico	p	10^{-12}
femto	f	10^{-15}
atto	a	10^{-18}
zepto	z	10^{-21}
yocto	y	10^{-24}

The SI unit of measure for the base quantity *mass* is the *kilogram*. It is the only unit which is still defined as a physical artefact; a platinium–iridium alloy prototype declared by the third general conference for weights and measures (1901) to be the unit of mass. It is kept at the international bureau for weights and measures (BIPM) in Sèvres, near Paris, France.

The SI unit of measure for the base quantity *time* is the *second*. It is defined as the duration of 9 192 631 770 cycles of electro-magnetic radiation corresponding to the transition between the two hyperfine levels of the ground state of the caesium-133 atom.

The SI unit of measure of the base quantity *electrical current* is the *ampere*. It is defined as the current which produces a force of 2×10^{-7} newton per metre of length between two long conductors placed in parallel one metre apart in vacuum.

The SI unit of measure for thermodynamic *temperature*, the *kelvin* is defined as 1/273.16 of the thermodynamic temperature of the triple point of water.

The SI unit of measure for *luminous intensity* is the *candela*. It is defined as the intensity in a given direction of a source emitting monochromatic radiation in that direction at a frequnecy of 540×10^{12} hertz with a radiant intensity of 1/683 watt per steradian.

The SI unit of measure for the *amount* of *substance* is the *mole*. One mole contains as many elementary items as there are atoms in 0.012 kilogram of carbon-12. (Here the lavbel 'elementary items' has to be specified. It may pertain to atoms, molecules, ions, electrons or to other particles or collections of identical entities.)

The definitions of the two additional dimensionless quantities are given as:

– The *radian* is the plane angle subtended at the centre of a circle by an arc equal in length to the radius of the circle.
– The *steradian* is the solid angle subtended at the centre of a sphere by an area on the surface of the sphere, which is equal to the square of the radius of the sphere.

A.2 Notation of measurement results

The outcome of a measurement or the result of a calculation based on measured values will be a numerical value with a given inaccuracy. The magnitude of the inaccuracy is a measure of the certainty with which the thus obtained value may be regarded. If a measurement result is stated without its inaccuracy, it does not provide any information and is thus without meaning. Therefore, a measurement result is generally given together with the maximum possible error. The result of an electrical current measurement may be, for instance, given as 6.35 ± 0.03 A or 6.35 A $\pm 5\%$.

If the error is not explicitly stated with the measurement result, it is tacitly agreed that the absolute error is equal to plus or minus one smallest decimal. A measurement result of 7.35 A therefore implies 7.35 ± 0.01 A. If, for example, a potentiometer is adjusted with a resolution of 1% to the value of $1\,000\,000$ Ω, this is noted as 1.000 MΩ or 1000 kΩ, since the notation $1\,000\,000$ Ω would suggest an error of no more than $\pm 10^{-6}$.

Obviously, this does not apply to exact mathematical constants (such as $\sqrt{2}$, 2π, π^2, etc.) which are often found in physical equations.

NB Many units are named after famous physicists of the past. In order to avoid confusion, it has been agreed that when the unit is written out in full, it must begin with a lower case letter. The abbreviation of the unit originating from a name will start with a capital. So, for instance, the SI unit for electrical current is named after André Marie Ampère (1775–1836), a French physicist pionering in electro-magnetism. A measured current would therefore be indicated as 10.5 ampere or 10.5 A.

When the results of a series of measurements are represented by a graph (for instance in an xy plot), often scaling factors are used in order to demonstrate that different experiments produce comparable results. This is usually indicated by labelling the axis 'arbitrary units'. However, the *units* are definitely not arbitrary; the *scaling factors* are arbitrary. So the text along the axis should read 'arbitrary scaling factor'.

Finally, we will make several remarks with respect to conventions concerning the nomenclature of measurement.

- Often, if an RMS-value of a given voltage is indicated, say 25 V, then this is noted as 25 V_{RMS}. However, this unjustly suggests that there are several different kinds of units: V and V_{RMS}. Therefore, a better notation is: 'The RMS-value of the voltage is 25 V'. This may also be written as: $V_{RMS} = 25$ V.
- Units are always used in their singular form, so it is incorrect to say: 'The potential amounts to 10 volts'.
- A temperature is expressed in SI units simply in kelvin and not in degrees kelvin. Likewise, we do not speak of divisions volt.
- Strictly speaking, the names voltmeter, ampere meter, etc. are incorrect. These names suggest the meters measure units and not quantities such as potential, current, etc. It would be more appropriate to use names such as potential meter, current meter, etc.

– It is logical for a physical characteristic to be related to a parameter value in such a way that as the characteristic becomes more pronounced, the parameter which quantifies this characteristic increases in value. For instance, a measurement system whose transfer function deviates only 1% from the desired true value is described by an inaccuracy of 1% (or, possibly, an accuracy of 99%), but not by an accuracy of 1%. In the same manner, one speaks of a non-linearity 10^{-3} but not a linearity of 10^{-3}. The resolution of a high-resolution measurement must be stated as 10^4 and not as 10^{-4}.

A.3 Decibel notation

The (deci)bel is a logarithmic measure for a power ratio. The power P_2 is *a bel* greater than P_1 when:

$$\log_{10}\left(\frac{P_2}{P_1}\right) = a$$

Thus, the power P_2 can be written as:

$$P_2 = 10^a P_1$$

The *decibel* (dB) is a tenth of a bel, so:

$$a \text{ bel} = 10a \text{ dB} = 10 \log_{10}\left(\frac{P_2}{P_1}\right) \text{dB}$$

It is also possible to use the (deci)bel as an absolute measure of power by assigning a fixed reference value P_r to P_1 and relating all subsequent power calculations to P_r. Most often $P_r = 1$ mW is chosen. In this case, the unit is denoted Bm or dBm. Therefore, a power of 0 dBm is equal to 1 mW, 10 dBm of power is the same as 10 mW, 20 dBm corresponds to 100 mW, etc.

The ratio of the amplitudes of two signals may also be expressed in decibels, for instance the ratio of two voltages or two currents. This is illustrated by the following example. An electronic system is supplied an input voltage V_i and has an input resistance of R_i, in which P_i watt is dissipated. The system produces an output voltage across a load resistance R_o, in which a power P_o is delivered. Remembering that $P = V^2/R$, we find for the power amplification of the system in decibel notation:

$$10 \log_{10}\left(\frac{P_o}{P_i}\right) = 20 \log_{10}\left(\frac{V_o}{V_i}\right) + 10 \log_{10}\left(\frac{R_i}{R_o}\right) \text{dB}$$

If, by coincidence $R_i = R_o$, the power amplification becomes equal to:

$$20 \log_{10}\left(\frac{V_o}{V_i}\right) \text{ dB}$$

This last expression is also often used as a logarithmic measure of the ratio of two amplitudes, even if the input resistance R_i is not equal to the output resistance R_o. The correction factor $10\log_{10}(R_i/R_o)$ is usually omitted for reasons of convenience. If this method is followed (as is general practice in electronics), then it is important to indicate clearly that this expression is used as an *amplitude ratio* instead of a *power ratio*, to avoid confusion.

Finally, the *neper* (Np) has been defined as the natural logarithm of the ratio of two amplitudes (potential, current, pressure, etc.). Hence, the voltage gain in neper is simply given by:

$$\ln\left(\frac{V_o}{V_i}\right) \text{ Np}$$

A.4 *V*- and *I*-quantities

In Section 2.2 we saw that the behaviour of non-electrical physical systems can often be described in the form of an electrical analogy. The behaviour of the physical system and that of the electrical circuit, which serves as the analogy, are described by identical mathematical expressions, if the physical quantities in the system can be related to corresponding quantities in the analogy. This relationship is generally called the equivalence relation.

We have to distinguish between two types of quantity: variables and parameters. The *variables* represent the energetic signal quantities, such as potential difference, pressure, velocity, etc. and the parameters represent the non-energetic system or network quantities, such as resistance, stiffness, mass, etc.

A classification which is especially suitable for measurement purposes is the division into across variables or transvariables and through variables or pervariables. A *transvariable* is measured between two different points of a physical system (with a measurement instrument which has a high input impedance). A *pervariable* is measured at a single point in the system (by inserting a measurement system with a low input impedance).

Both the transvariables and the pervariables can be subdivided into *rate variables* and *state variables*. For instance, the electrical current i and the electrical charge q are both examples of pervariables, but the current i is a rate variable and the charge q is a state variable. The relation between i and q is given by $i = dq/dt$. This applies in general: the rate variable is the time derivative of the associated state variable. Thus, we are permitted to limit ourselves to discussing rate variables only. In Section 2.2, we called the rate transvariable a *V-quantity* and the rate pervariable a *I-quantity*, analogous to the usage in the electrical domain of physics.

If we restrict ourselves merely to *V*- and *I*-quantities, we can find equivalence relations for the quantities from various domains of physics, as indicated in Table A.5.

Analogous qunatities are listed in the columns for each domain of physics (horizontal entry). It should be noted that the mechanical impedance in this analogy gives the reciprocal of the prevailing definition of impedance in the field of mechanics. Furthermore, the heat analogy is, in fact, a pseudo-analogy. It would be more correct to replace the heat flux I_w by the entropy flux. The result of this pseudo-analogy is that the product of the *V*- and *I*-quantities is not power, as is the case for all other analogies. Consequently, the heat resistance is not a dissipating resistance. Finally, it should be remarked that a thermal self-inductance does not exist, and therefore, it is impossible to construct a thermal resonant circuit with passive components only.

Table A.5. *Analogy between variables and parameters of different domains of physics.* (U = *potential difference, I = current, R = resistance, C = capacitance, L = self-inductance, v = velocity, F = force, D_t = translational damping, m = mass, K_t = translational stiffness, ω = angular velocity, M = moment, D_r = rotational damping, J = moment of inertia, K_r = rotational stiffness, ΔT = temperature difference, I_w = heat flux, R_w = heat resistance, C_w = heat capacitance, Δp = pressure difference, I_v = rate of flow, R_s = flow resistance, C_s = flow capacitance, L_s = flow inertia.*)

Physical domain	Variables		Parameters			
electrical	V	I	$\overline{V/I}$	R	C	L
mechanical						
translation	v	F	$\overline{v/F}$	$1/D_t$	m	$1/K_t$
rotation	ω	M	$\overline{\omega/M}$	$1/D_r$	J	$1/K_r$
thermal (pseudo)	ΔT	I_w	$\overline{\Delta T/I_w}$	R_w	C_w	—
hydraulic ⎫ pneumatic ⎬ acoustic ⎭	Δp	I_v	$\overline{\Delta p/I_v}$	R_s	C_s	L_s

A.5 Tables

In this section, we will provide values for a number of physical constants which frequently occur in measurements. We will also list several conversion factors for non-SI units which, alas, are still often used. Finally, the electrical, mechanical and thermal properties of various materials which we regularly encounter are given.

A.5.1 Physical constants

–	Velocity of light in vacuum	$(2.99792458 \pm 0.00000003) \times 10^8$ m/s
		$= 1 / \sqrt{\mu_0 \varepsilon_0}$
–	Permeability in free space	$4\pi \times 10^{-7}$ H/m
–	Permittivity in free space	$(8.85416 \pm 0.00003) \times 10^{-12}$ F/m
–	Electron, charge	$(1.60217333 \pm 0.0000003) \times 10^{-19}$ C
	rest mass	$(9.1091 \pm 0.0004) \times 10^{-31}$ kg
	radius	$(2.81777 \pm 0.00005) \times 10^{-15}$ m
–	Proton, rest mass	$(1.67252 \pm 0.00008) \times 10^{-27}$ kg
–	Neutron, rest mass	$(1.67482 \pm 0.00008) \times 10^{-27}$ kg
–	Boltzmann constant	$(1.3804 \pm 0.0001) \times 10^{-23}$ J/K
–	Stefan–Boltzmann constant	$(5.6686 \pm 0.0005) \times 10^{-8}$ J/m^2 s K^4
–	Planck constant	$(6.6252 \pm 0.0005) \times 10^{-34}$ J s
–	Avogadro constant	$(6.02252 \pm 0.00003) \times 10^{23}$ mol^{-1}
–	Gas constant	8.314 ± 0.001 J/mol K
–	Josephson constant	4.834599 ± 0.0000004 GHz/V
–	Von Klitzing constant	25.812807 ± 0.0000002 kΩ
–	Standard acceleration of gravity	9.8067 ± 0.0001 m/s^2
–	Gravitational constant	$(66.7323 \pm 0.0001) \times 10^{12}$ N m^2/kg
–	Mass–energy conversion factor	$(8.98755 \pm 0.00003) \times 10^{16}$ m^2/s^2
–	Wave impedance (vacuum)	376.7Ω

A.5.2 Conversion factors

1 inch (1")	=	2.540×10^{-2} m
1 foot (1')	=	30.48×10^{-2} m
1 yard (1 yd)	=	0.9144 m
1 angstrom	=	10^{-10} m
1 mile (UK nautical)	=	1853 m
1 mile (statute)	=	1609 m

1 ounce (oz)	$= 28.35 \times 10^{-3}$ kg
1 pound (lb)	$= 453.0 \times 10^{-3}$ kg
1 carat	$= 200 \times 10^{-6}$ kg
1 dyne	$= 10^{-5}$ N
1 bar	$= 10$ N/m^2
1 atm	$= 1.013 \times 10^5$ N/m^2
1 cm Hg	$= 10$ torr $= 1.333 \times 10^3$ N/m^2
1 calorie	$= 4.187$ J

(A frequently used unit in physiology is the so-called large calorie, which is equal to 1 kilocalorie.)

1 BTU (British thermal unit)	
	$= 1059.52$ J
1 erg	$= 10^{-7}$ J
1 HP	$= 735.5$ W
1 oersted	$= (1/4\pi) \times 10^3$ A/m
1 maxwell	$= 10^{-8}$ Wb
1 gauss	$= 10^{-4}$ T
1 neper	$= 20 \log e$ dB $= 8.686$ dB
$(y\,°F - 32)\frac{5}{9}$	$= x\,°C$
1 foot-candle	$= 10.764$ lux
1 lambert	$= 10^4$ lux

A.5.3 Properties of materials

All properties are rated at a temperature of 20 °C.

The resistivity ρ and temperature coefficient $C_{T\rho}$ of several *conductors*.

Material	$\rho \times 10^{-8}\,\Omega\,\mathrm{m}$	$C_{T\rho} \times 10^{-3}$/K
silver	1.50	4.10
copper	1.55	4.33
gold	2.04	3.98
aluminium	2.50	4.67
brass	6–8	1–2
steel	8–14	6–7
manganin	43	< 0.005
constantan	49	< 0.04

The relative permittivity ε_r, loss angle tan δ, resistivity ρ and the breakdown field-strength E_d of a number of isolators.

Material	ε_r	$\tan \delta \times 10^4$		ρ	E_d
	(50 Hz)	(50 Hz)	(1 MHz)	$\times 10^{10}\,\Omega\,m$	$\times 10^6$ V/m
glass	5–9	5–30	1–10	> 10	100–400
rubber	3–5	20–100	50–150	10^5–10^6	20–30
porcelain	6	10–30	3–15	10^2–10^5	300–400
mica	4	5	1	> 10^3	100
oil	2–2.5	1–5	1–5	10^3–10^4	100–250
epoxy resin	3–4	30–50	100–200	10^3–10^4	80–120
Plexiglass	4	500	150	10^4	90
Teflon	2	< 1	< 1	10^5	250
PVC	4–6	100–1000	50–500	10–10^4	80

It follows from the above tables that the (free) charge relaxation time is $T = \varepsilon_0 \varepsilon_r \rho$. This is the time required for the field strength in a conductor to return to zero, so $q(t) = q_0 \exp(-t/T)$ when all free charge has re-distributed itself, after a point charge q_0 has been injected into the conductor.

The density ρ_m, Curie temperature T_c, relative initial permeability μ_{ra} and the resistivity ρ of a number of *soft magnetic materials*.

Material	ρ_m	T_c	μ_{ra}	ρ
	$\times 10^3$ kg/m^3	°C	$\times 10^3$	$\times 10^{-8}\,\Omega\,m$
iron	7.8	770	0.3–1	10
silicon iron	7.6	750	0.3–3	40–55
nickel iron	8.5	250–450	2–130	40–75
ferrites	4.5	130–400	0.3–10	10^7–10^{13}

The density ρ_m, coefficient of elasticity E_m and related temperature coefficient C_{TE} and the modulus of rigidity G_m and related temperature coefficient C_{TG} of several frequently used *construction materials*.

Material	ρ_m	E_m	C_{TE}	G_m	C_{TG}
	$\times 10^3$kg/m^3	$\times 10^{10}$ N/m^2	$\times 10^{-4}$ /K	$\times 10^{10}$ N/m^2	$\times 10^{-4}$ /K
α-iron	7.86	21	2.5	8.2	3.0
aluminium	2.70	7.1	4.4	2.7	4.9
copper	8.92	12	3.9	4.6	4.0

The density ρ_m. heat resistivity r, specific heat capacitance c, linear coefficient of expansion α and the coefficient of emission ε of several other *construction materials*.

Material	ρ_m $\times 10^3$ kg/m^3	r $\times 10^{-3}$ mK/W	c $\times 10^3$ J/kg K	α $\times 10^6$ /K	ε
aluminium	2.70	4.90	0.896	23.8	0.04–0.08
copper	8.92	2.59	0.383	16.8	0.04–0.05
iron	7.86	13.7	0.452	11.0	0.08
Invar	8.14	93.5	0.460	1.2	0.07
glass	2.51	1320	0.745	3.2–4.2	0.88
PVC	1.2–1.5	5800–6250	1.3–2.1	150–200	0.90
Plexiglass	1.18	5260–6250	1.2–1.4	70–100	0.85
Teflon	2.1–2.3	4400–6250	0.9–1.1	60–100	0.90

Bibliography

Theory of measurement
Suppes, P., Zinnes, J.L., 'Basic measurement theory', *Handbook of mathematical psychology*, John Wiley (London, 1953).
Ellis, S., *Basic concepts of measurement*, Cambridge University Press (Cambridge, 1966).
Krantz, D.H., Luce, R.D., Suppes, P., Tversky, A., *Foundations of measurement*, Vol. 1, Academic Press (New York, 1971).
Sydenham, P.H., *Handbook of measurement science*, John Wiley (New York, 1982).

Transducers
Doebelin, E.O., *Measurement Systems*, Chapter 2: 'Measuring devices', pp. 212–604, MacGraw-Hill (New York, 1966).
Norton, N.H., *Handbook of transducers for electronic measurement systems*, Prentice Hall (Englewood Cliffs, 1969).
Nelting, H., Thiele, G., *Elektronisches Messen nichtelektrischer Grössen*, Philips Technische Bibliotheek (Eindhoven, 1966).
Rohrbach, C., *Handbuch für elektrisches Messen mechanischer Grössen*, V.D.I.-Verlag (Düsseldorf, 1967).
Wieder, H.H., *Hall generators and magnetoresistors*, Pion Ltd (London, 1971).

Electronic measurement devices
Harris, F.K., *Electrical measurements*, John Wiley (New York, 1956).
Hague, S., Foord, T.R., *Alternating current bridge-methods*, Pitman Press (London, 1971).
Jenkins, B.D., *Introduction to instrument transformers*, George Newnes Ltd (London, 1967).
Stout, M.B., *Basic electrical measurements*, Prentice Hall (Englewood Cliffs, 1962).

Frank, E., *Electrical measurement analysis*, MacGraw-Hill (New York, 1959).

AD and DA conversion
Schmid, H., *Electronic analog–digital conversions*, Van Nostrand Reinhold (New York, 1970).
Sheingold, D.H., *Analog–digital conversion handbook*, Analog Devices (Norwood, Mass. USA, 1972).

Electronic measurement systems
Coombs, C.F., *Basic electronic instruments handbook*, MacGraw-Hill (New York, 1972).
Oliver, B.M., Cage, J.M., *Electronic measurements and instrumentation*, MacGraw-Hill (New York, 1971).
Prensky, S.D., *Electronic instrumentation*, Prentice Hall (Englewood Cliffs, 1971).

Journals
The Review of Scientific Instruments (USA).
Journal of Scientific Instruments (GB).
IEEE Transactions on Instrumentation and Measurements (USA).
Archiv für technisches Messen (Germany).

Index